The Arctic in China's National ~~~

This book locates the Arctic within the context of the People's Republic of China's (PRC) national strategy of the Great Rejuvenation of the Chinese Nation. Drawing on a range of sources published in Chinese and English, the author analyses Beijing's Arctic scientific activities and technological capabilities, including the research infrastructure, long-term goals, and the significance for China's understanding of the region, its Arctic identity, and international perceptions. Examining the region from the perspective of the Comprehensive National Security Outlook developed during the Xi Jinping era, the book focuses on military, economic, technological, and political components and considers the PRC's official and academic discourses regarding Arctic governance and the views of the region within bilateral relations with Arctic states, outlining a science, security, and governance nexus in China's Arctic engagement.

This volume will be of interest to scholars and students of Arctic geopolitics, Chinese studies, security studies, and foreign policy analysis. It will also appeal to policymakers and defence analysts in Arctic states and other regional stakeholders.

Martin Kossa is an Associate Professor at Nord University in Norway. He received his PhD from City University of Hong Kong.

Rethinking Asia and International Relations

Series Editor – Emilian Kavalski, Li Dak Sum Chair
Professor in China-Eurasia Relations and International Studies,
University of Nottingham, Ningbo, China

This series seeks to provide thoughtful consideration both of the growing prominence of Asian actors on the global stage and the changes in the study and practice of world affairs that they provoke. It intends to offer a comprehensive parallel assessment of the full spectrum of Asian states, organisations, and regions and their impact on the dynamics of global politics.

The series seeks to encourage conversation on:

- what rules, norms, and strategic cultures are likely to dominate international life in the 'Asian Century';
- how will global problems be reframed and addressed by a 'rising Asia';
- which institutions, actors, and states are likely to provide leadership during such 'shifts to the East';
- whether there is something distinctly 'Asian' about the emerging patterns of global politics.

Such comprehensive engagement not only aims to offer a critical assessment of the actual and prospective roles of Asian actors, but also seeks to rethink the concepts, practices, and frameworks of analysis of world politics.

This series invites proposals for interdisciplinary research monographs undertaking comparative studies of Asian actors and their impact on the current patterns and likely future trajectories of international relations. Furthermore, it offers a platform for pioneering explorations of the ongoing transformations in global politics as a result of Asia's increasing centrality to the patterns and practices of world affairs.

Recent titles

Chinese Paradiplomacy at the Peripheries
Beyond the Hinterland
Yao Song and Tianyang Liu

The Arctic in China's National Strategy
Science, Security, and Governance
Martin Kossa

For more information about this series, please visit: www.routledge.com/Rethinking-Asia-and-International-Relations/book-series/ASHSER1384

The Arctic in China's National Strategy

Science, Security, and Governance

Martin Kossa

Routledge
Taylor & Francis Group

LONDON AND NEW YORK

First published 2024
by Routledge
4 Park Square, Milton Park, Abingdon, Oxon OX14 4RN

and by Routledge
605 Third Avenue, New York, NY 10158

Routledge is an imprint of the Taylor & Francis Group, an informa business

British Library Cataloguing-in-Publication Data
A catalogue record for this book is available from the British Library

ISBN: 978-1-032-27948-0 (hbk)
ISBN: 978-1-032-28053-0 (pbk)
ISBN: 978-1-003-29511-2 (ebk)

DOI: 10.4324/9781003295112

Typeset in Times New Roman
by Apex CoVantage, LLC

Mojim rodičom

Contents

Acknowledgements

I would like to thank a few people whose encouragement, support, and insights contributed to the development and completion of this book.

My sincere gratitude goes to my PhD supervisor, Nicholas Thomas. This book is based on my PhD research at City University of Hong Kong concluded in 2019. Without Nick's expertise, advice, and care, which guided my doctoral dissertation and shaped my professional development, this book would not be possible.

I am also grateful to my colleagues at Nord University in Norway. Beate Steinveg and Torbjørn Pedersen read parts of the manuscript and provided valuable feedback. They also engaged in conversations with me about Arctic geopolitics and governance as seen from the Norwegian perspective. Because of their feedback and our conversations, I was able to improve the analysis in the manuscript. Corine Wood-Donnelly commented on the initial book proposal and was always available when I appeared in her office and needed advice on overcoming hurdles associated with writing a book.

I owe a debt of gratitude to Iselin Stensdal from the Fridtjof Nansen Institute in Norway. Iselin was very kind to comment on parts of the manuscript and invite me to two workshops in Oslo in 2021 and 2022, where I was able to present the main ideas and arguments of the book. Moreover, I am thankful to the two anonymous reviewers of the book proposal for their constructive comments, and the editorial team at Routledge for their patience and advice. Any errors are my sole responsibility.

My greatest thanks go to my partner, Kerri, whose love, support, and beautiful smile kept me going during the cold days in the Norwegian Arctic. Kerri, I am very fortunate to have you in my life.

This book is dedicated to my parents for their love and constant support that allowed me to pursue my ambitions.

1 Introduction

On October 10, 2017, after 83 days at sea and 20,000 nautical miles, the Chinese research icebreaker, the Xuelong, returned to Shanghai and successfully completed the People's Republic of China's (PRC) eight Arctic scientific expedition. During this expedition, Chinese scientists conducted a comprehensive survey of the Arctic marine environment, including observations of Arctic sea ice, biodiversity, ocean acidification, and marine microplastic accumulation, thus continuing the tradition of China's scientific expeditions to the region, which officially commenced in the late 1990s. It was, however, also during this expedition that the Xuelong accomplished a tremendous feat by circumnavigating the Arctic Ocean. The Xuelong managed to sail toward Europe by traversing the high-latitude Transpolar Sea Route (TSR) cutting across the central part of the Arctic Ocean. On its return journey, it sailed through the Northwest Passage (NWP) across the Canadian archipelago.[1] These sea routes harbour great potential for commercial shipping by reducing the distance and travel time between North America, Europe, and East Asia. By traversing these shipping routes, Chinese scientists accumulated a wealth of Arctic meteorological, sea ice, and seabed data from along these waterways. This contributed to the development of a Chinese marine environment numerical forecasting system and advanced the PRC's systematic assessment of the navigability of Arctic shipping lanes.[2]

Since then, the PRC has published a dedicated Arctic White Paper which identified China as an important Arctic stakeholder, outlined its regional interests, and positioned China geographically as a near-Arctic state (近北极国家). The country formally opened a second (joint) Arctic research station in Iceland and launched a second, domestically constructed research icebreaker, the Xuelong 2. At the same time, Chinese state-owned enterprises (SOE) are actively considering how to further expand their presence in the region (especially in areas of resource development and shipping), capitalising on their successful involvement in the Yamal Liquefied Natural Gas (LNG) project in the Russian Arctic zone and commercial shipments on the Northern Sea Route (NSR). The Arctic has also emerged as a new dimension of the expanding Sino-Russian partnership. Navies of both states conduct drills in sub-Arctic waters, at times causing alarm in the capitals of Western Arctic states. More recently, the Russian and Chinese coastguards signed an agreement regarding cooperation in law enforcement at sea which will see Chinese

DOI: 10.4324/9781003295112-1

participation in joint maritime exercises in the Russian Arctic in the near future.[3] Such developments, however, imply that the PRC seems to consider the Arctic from a more strategic and utilitarian perspective that goes beyond scientific observations of the Arctic environment.

An increasing Chinese scientific, economic, and strategic focus on the Arctic has emerged amid substantial shifts in how the PRC structures its economy and armed forces and, more importantly, how it perceives and re-evaluates its position within the hierarchy of the great powers of the international system. These shifts began after the 2008 financial crisis, but they received a significant boost in confidence, depth, and assertiveness when Xi Jinping assumed the role of General Secretary of the Chinese Communist Party (CCP) in late 2012. Under his leadership, the *Chinese Dream of the Great Rejuvenation of the Chinese Nation* became the PRC's national strategy, marking a new phase in China's development aimed at achieving socio-economic modernity and expanding national power. In 2017, at the 19th CCP Congress, shortly after the Xuelong completed its eight Arctic expedition, General Secretary Xi proclaimed that the PRC had entered a New Era, where China would strive to become a global leader with significant international influence.[4] To achieve this, the PRC has been vigorously pursuing an economic development model that relies on innovation and which is underpinned by a strong focus on science and technology. An equally important part of Xi's national strategy relates to security, encompassing both internal and external dimensions and embodied in the concept of the Comprehensive National Security Outlook. This outlook also includes what the Chinese labelled as the strategic new frontiers (战略新疆域), namely, the polar regions, the deep sea, outer space, and cyberspace. Xi Jinping's China is defined by a much more repressive state apparatus that suppresses any form of dissent through its vast surveillance capabilities. Additionally, through the adoption of various laws, the scope of national security has expanded significantly. This has been accompanied by increasing military budgets that the People's Liberation Army (PLA) uses to develop and acquire advanced weapon systems and platforms, such as fifth-generation stealth fighter jets, aircraft carriers, and ballistic missiles. Xi is also confident that a rejuvenating China will be able to provide the world with solutions, technologies, and capital that could resolve the perceived gaps in global governance (in security and development, for example) that current leading great powers are unable or unwilling to address. CCP leadership seems to be convinced that Chinese style development is a new form of modernisation that other countries, especially in the developing world, can adopt.[5] China's Belt and Road Initiative (BRI) is being developed as a vehicle through which this can be delivered and through which the PRC further links itself with neighbouring countries and distant regions alike via infrastructure development, trade, finances, and cultural exchanges. To paraphrase Shue, it appears that when Xi and other CCP leaders discuss the PRC's rejuvenation and their aspirations for China, they draw inspiration from the country's imperial past. During that period, China held the position of being the dominant economic, military, and cultural power in the region. It served as a central hub for trade, learning, and advanced production methods earning admiration and deference from others.[6]

This book attempts to locate the Arctic and what the region has to offer, such as its strategic location, scientific knowledge, natural resources, shipping lanes, and role in regional governance, within the context of the PRC's national strategy of the Great Rejuvenation of the Chinese Nation as seen from the Chinese perspective and portrayed in official PRC policy documents, as well as Chinese elite discourses. Such an undertaking is particularly relevant as the PRC, under the leadership of General Secretary Xi, endeavours to reposition itself as a global science, technology, innovation, and high-tech manufacturing hub, which actively promotes its own visions of the global order and the arrangements of various subregional groupings and China's role in them. The PRC of today seems to possess the will, resources, expanding capabilities, and diplomatic instruments to shape the international order to its preferences in order to implement its vision of a future global order. This comes at a time when China's relations with the United States (US) are deteriorating at an exponential rate. The two nations find themselves engaged in a power struggle that extends far beyond the Indo-Pacific region, spanning multiple domains including space, technology, trade, development, security, finances, and even global norms in areas such as human rights. From a regional perspective, the PRC's considerable economic influence, the changing perceptions of its position in the global and regional orders, the multilayered competition with the US and the close cooperative partnership with Russia will also have implications for Arctic regional affairs. This is especially true in areas of maritime security, infrastructure, resource development, and Arctic governance. In addition, as the Arctic seems to attract much extra-regional attention, the region could be considered a microcosm of interactions between established and emerging great powers that foreshadows power shifts and rearrangements at the global level.[7] All of this makes it even more important to understand how and in what capacity the PRC views the Arctic region.

This chapter is composed of four parts. The first part introduces the general conditions of the Arctic region. These include the state of Arctic environmental changes, the economic and commercial potential of the region, and the growing great power competition. The second part outlines the broad contours of the PRC's engagement with the Arctic since the early 1990s when China began to be more substantially interested in the region through scientific research. Part 3 notes the Chinese and English language sources used in the process of writing the book, and Part 4 outlines the organisation of the book.

The Evolving Arctic

The international profile of the Arctic region[8] has been growing steadily over the past three decades. This is primarily due to three reasons. First, the Arctic is experiencing changes in its biophysical environment that have local as well as global impacts. This is related to rising regional temperatures. The increase in annual Arctic mean land and ocean surface temperature is now almost three times higher than the increase in the global average.[9] As a result, Arctic sea ice, snow cover, glaciers, and permafrost have undergone reductions in size and thickness. For example, in 2012, the Arctic sea ice extent reached 3.41 million km², the lowest seasonal minimum extent in the satellite record since 1979. According to the National Snow and Ice Data Centre at the University of Colorado Boulder, the lowest 16 sea ice extents in the satellite record were observed in the past 16 years, from 2007 to 2022.[10] These changes lead to or can accelerate existing issues including coastal erosion, boreal forest and tundra wildfires, inland flooding, extreme temperature, and precipitation events, as well as disruptions to infrastructure built on the now melting permafrost.[11] Beyond these, Arctic environmental changes seem to be impacting regions further south in the temperate zone of the Northern Hemisphere. Some members of the international scientific community have suggested that the melting of the Greenlandic ice cap (and other glaciers) is contributing to sea level rise, that melting permafrost is releasing methane (a greenhouse gas) into the atmosphere, thereby contributing to global temperature increases, and that changes in the Arctic could be linked to mid-latitude extreme weather events such as droughts.[12] Given these changes, international scientific communities and states are devoting considerable resources and manpower to understanding the impacts of Arctic environmental changes. For example, Japan has launched a national flagship project for Arctic research, the Arctic Challenge for Sustainability II, to study the status and process of Arctic environmental changes and assess their impacts on human society.[13] At the international level, hundreds of researchers from 20 nations launched the MOSAiC (Multidisciplinary Drifting Observatory for the Study of Arctic Climate) expedition in September 2019 aboard the German research vessel Polarstern into the central Arctic to explore the Arctic climate system and its role in global climate change.[14]

Second, these environmental changes add to global perceptions of the Arctic as a region that is opening up and whose resources and maritime routes are ready for commercial exploitation. The Arctic shows considerable potential for the mining of various minerals, metals, and gems, including rare-earth elements, coal, iron ore, nickel, cobalt, zinc, copper, gold, silver, platinum, and diamonds. In addition, parts of the Arctic region are basins that contain exploitable quantities of oil and natural gas, some of which are already in production, such as the oil fields in Alaska and Siberia. The perception of the Arctic as a source of hydrocarbons was further accentuated by the publication of geological surveys which estimated large amounts of undiscovered oil and gas reserves located above the Arctic Circle.[15] In terms of shipping, three shipping routes in the Arctic Ocean are being considered due to their potential to cut distances and shipping times between the Northeast Asian, European, and North American markets. These include the aforementioned TSR across the Central Arctic Ocean, the NWP along the northern coast of Alaska and through the Canadian Archipelago, and the NSR along the Russian coastline. Due to sea ice conditions and state policy attention, the NSR has the greatest potential for development.[16] The economic potential of these shipping routes is being promoted through various assessments and by Arctic leaders. Russia's President Putin has previously stated that Arctic shipping lanes are almost a third shorter than the lanes passing through the Suez Canal and predicted that the NSR "will rival traditional trade lanes in service fees, security, and quality."[17] Consequently, the economic possibilities of the Arctic and the perception of its exploitable potential have captured global interest from emerging economies, leading to their inclusion in the official Arctic policy documents of these states. For example, India's 2022 Arctic policy outlines one of its objectives as being to "explore opportunities for responsible exploration of natural resources and minerals in the Arctic."[18]

Third, the Arctic region is becoming an arena for great power rivalry, following a period of relative regional calm in the 1990s and early 2000s. The Russian state is reaffirming its presence in the region and asserting its status as an Arctic great power through various means, including the retrofitting and expansion of its Arctic military bases, strengthening of strategic forces, fortification of coastal defences, and enhancement of naval capabilities.[19] Similarly, the US is upgrading its military installations in Alaska and Greenland, while also expanding its presence in the European Arctic. For instance, it has redeployed forces to its Cold War–era bases in Iceland. Additionally, the navies of non-Arctic nations like the UK have been actively operating in the Barents Sea in recent years, and European allies' armies are conducting military exercises and cold-weather training in northern parts of Scandinavia. Simultaneously, the Arctic is not exempt from the repercussions of geopolitical, military, and geo-economic conflicts at the global level. Russian actions targeting Ukraine in both 2014 and 2022 resulted in the breakdown of cooperative mechanisms previously established among Arctic states, which focused on environmental protection, sustainable development, and soft security initiatives like coast guard exchanges. This disruption effectively halted progress in Arctic governance. To address this void, multilateral platforms, including international knowledge-sharing events and conferences that can accommodate global actors, have gained popularity as avenues for discussing regional matters.[20] In addition, as a result of international sanctions imposed on Russia in response to its invasions of Ukraine, the Kremlin was compelled to re-evaluate its stance on external involvement in its Arctic zone. Consequently, Russia has become more open to foreign investors, particularly those from Asia. This shift is generating additional opportunities for extra-regional great powers, such as China, to express their ideas and policy preferences regarding the region and to invest in Arctic projects.

The PRC and the Arctic Since the 1990s

Although Chinese policy documents, analysts, and academics highlight that China's Arctic engagement can be traced back to 1925 when the country joined the Svalbard Treaty, thus legitimising its presence in the Arctic,[21] the substantive involvement of China in the region truly began in the 1990s. During this time, Chinese scientists started participating in collaborative regional projects. For instance, in 1991, a scientist from the Chinese Academy of Sciences (CAS) joined an Arctic expedition near Svalbard, collaborating with scientists from Norway, the Soviet Union, and Iceland.[22] This development can partially be attributed to the evolving structure of the international system. With the disintegration of the Soviet Union and the waning of the intense bipolar competition characterising the Cold War era, the security environment in the Arctic also underwent changes, leading the region to emerge as a potential platform for international cooperation.[23] Chinese national scientific associations, led by the China Association for Science and Technology (CAST)[24] played a crucial role in China's early scientific involvement in the Arctic.[25] CAST assisted in preparing for the first, albeit semi-official, Chinese expedition to the Arctic. On May 6, 1995, a group of seven Chinese scientists and reporters successfully reached the North Pole, becoming the first mainland Chinese to do so in the post–Cold War era.[26] Building on this milestone, in 1996, the PRC became the 16th member of the International Arctic Science Committee (IASC), a non-governmental organisation that facilitates multidisciplinary research in the Arctic region.[27]

In 1999, the PRC launched its first government-sanctioned scientific expedition to the Arctic, utilising the research icebreaker Xuelong. Spanning 71 days, the expedition covered the Bering, Chukchi, and Beaufort Seas and involved scientists from various research institutes across China, accompanied by several reporters.[28] Following these initial successes, some in the Chinese scientific community advocated for the establishment of a dedicated Chinese research station in the Arctic. Once again, Chinese scientific associations took the lead. With support from two corporations, the Xinjiang Yilite Industrial Company and Hunan Mulin Food Company, the China Association for Scientific Expedition established a temporary *Chinese Arctic Yilite-Mulin Research Exploration Station* (中国伊力特＊沐林北极科学探险考察站) in Longyearbyen, Svalbard, in July 2002, with the approval of Norway.[29] The purpose of this station was to facilitate a comparative study of the ecological environment between Svalbard and the Qinghai-Tibet Plateau and pave the way for the Chinese government to establish a long-term Arctic research station in Svalbard.[30] It remained active until September 2003. In light of these developments, the PRC's State Oceanic Administration (SOA), responsible for polar exploration at the time, contemplated establishing a permanent presence in the Arctic. PRC officials initiated negotiations with Norwegian state representatives, and by June 2003, the SOA had signed a lease agreement for a Chinese research station on Svalbard with the Kings Bay company, the primary provider of facilities and logistics for researchers on Svalbard.[31] Consequently, in July 2004, China inaugurated its first official Arctic research station, named the *Arctic Yellow River Station*, located in Ny-Ålesund, Svalbard.

At this time, Gunnar Palsson, an Icelander who had served as the Chairman of the Senior Arctic Officials of the Arctic Council (AC), a prominent regional governance forum, visited Beijing and met with representatives from the Chinese Ministry of Foreign Affairs (MFA). China was back then the second largest emitter of CO_2 globally, and the urgency to engage it in matters related to Arctic climate change was significant.[32] Following Palsson's visit, the PRC formally applied to become an observer to the AC at the end of 2006. However, there is no concrete evidence that would prove that China's AC application was a direct result of Palsson's visit.[33] Since then, China has participated in AC activities on an *ad hoc* basis, requiring formal invitations from regular AC members to attend meetings.

China's perceptions of the Arctic as a region of geopolitical and economic significance underwent a shift in the late 2000s. The competing sovereignty claims among Arctic states gained attention after a Russian expedition planted a titanium

Russian flag onto the Arctic seabed at the North Pole.[34] Other factors contributing to this shift included the increasing viability of Arctic shipping routes due to record-low sea ice extents and the 2008 publication of the US Geological Survey's (USGS) assessment of undiscovered oil and gas resources above the Arctic Circle. This assessment indicated that substantial hydrocarbon deposits remain to be found in the Arctic.[35] As a result, China began assigning political significance to the Arctic region. In 2010, Liu Zhenmin, then Chinese Assistant Minister of Foreign Affairs, participated in the High North Study Tour organised by the Norwegian government, where he presented China's views on Arctic affairs. Liu emphasised that China was influenced by the evolving Arctic dynamics, advocated for scientific cooperation as the foundation for Arctic collaboration, and stressed the need for cooperation between Arctic and non-Arctic states to address transregional challenges facing the Arctic. He also highlighted the emerging economic opportunities in shipping due to melting sea ice.[36] In parallel, Chinese academic institutions expanded their interest in the Arctic, with the Ocean University of China and Dalian Maritime University establishing Arctic-focused research centres. In 2010, these universities, along with the Polar Research Institute of China (PRIC), the National Marine Environment Forecasting Centre, and COSCO (a state-owned enterprise) conducted a study that focused on the emerging Arctic shipping lanes.[37]

The PRC also actively engaged in bilateral diplomacy with Arctic states. In April 2012, then Chinese Premier Wen Jiabao made a historic visit to Iceland, becoming the first high-ranking CCP official to do so in decades. During the visit, both parties signed a Framework Agreement on Arctic cooperation.[38] Subsequently, the PRC embarked on its fifth Arctic scientific expedition, with the research icebreaker Xuelong sailing to Iceland via the NSR.[39] These developments, coupled with the growing recognition by the CCP leadership of the influence of Arctic geography and changing climate conditions on the Chinese mainland, led Chinese officials and specialists to begin referring to China as a near-Arctic state in 2012.[40] This designation signalled China's willingness to expand its participation in regional affairs to the international community.

The Xi Jinping Era

Under Xi's leadership, the PRC continued to expand its engagement with the Arctic, solidifying its interests and presence in the region. A significant milestone occurred in 2013 when China, along with Japan, South Korea, Singapore, India, and Italy, was granted formal observer status at the AC. This provided the PRC with unhindered access to AC meetings, rationalised its presence in the region, and contributed to a view of China as a regional stakeholder.[41] In the realm of science, the PRC continued to launch regular expeditions to the Arctic and enhance international scientific exchanges with Arctic states. For example, in December 2013, Chinese and Nordic research institutes established the China-Nordic Arctic Research Centre (CNARC 中国-北欧北极研究中心) in Shanghai as a collaborative platform focusing on Arctic politico-economic issues and sustainable development.[42] Furthermore, in 2016, the SOA signed a Memorandum of Understanding

(MoU) with Greenland on Arctic scientific cooperation, which included, among other things, discussions on establishing a Chinese research station on the island.[43] Science and science diplomacy were increasingly seen in China as tools that could assist the Party-state to comprehend the region's environmental changes and enhance its participation in regional affairs and governance.[44] Simultaneously, Chinese SOEs started exploring the economic potential of the opening Arctic. In August and September 2013, the cargo vessel Yong Sheng owned by COSCO, the largest shipping SOE in China, successfully sailed from Dalian in China's northeast to Rotterdam via the NSR, saving nine days in transit time compared to the Suez Canal route and becoming the first Chinese commercial vessel to achieve this.[45] Additionally, in 2014, the China National Petroleum Corporation (CNPC), a major Chinese oil and gas SOE, acquired a 20% stake in the Yamal LNG project in the Russian Arctic. Other Chinese companies also expressed intentions to develop mining projects in Greenland.

To highlight the increasing significance of the Arctic in the PRC's scientific, economic, and diplomatic endeavours, the Party-state leadership appointed a Special Representative for Arctic Affairs and initiated the development of the Polar Silk Road (PSR) through the Arctic Ocean, building on Sino-Russian regional economic cooperation. This development effectively introduced a third route within Xi's signature foreign policy initiative, the BRI, alongside the Silk Road Economic Belt and the 21st Century Maritime Silk Road. However, the high point of the PRC's Arctic engagement came with the release of its Arctic White Paper in January 2018. This policy document presented a vision of a changing Arctic that affects both regional and extra-regional states, with implications for the international community as a whole. It acknowledged that extra-regional states did not possess sovereign rights in the Arctic but highlighted their entitlement to certain rights under international law. These rights encompass scientific research, navigation, overflight in "relevant sea areas in the Arctic Ocean," and resource exploration and exploitation in areas beyond the national jurisdictions of Arctic states. Geographically, the white paper underlined China's proximity to the region, identifying the country as a near-Arctic state and an important stakeholder in Arctic affairs. It recognised that China's climate system, ecological environment, and by extension, its economic interests in agriculture and marine industry are directly influenced by Arctic changes. The policy paper also underscored that China, as a major trading nation and energy consumer, would be profoundly impacted by the opening of new Arctic shipping routes and the development of Arctic natural resources. The document outlined four policy goals: (1) understanding the Arctic; (2) protecting the Arctic; (3) developing the Arctic; and (4) participating in Arctic governance. These goals were to be achieved based on the principles of mutual and reciprocal respect, multilevel cooperation, win-win outcomes, and sustainability. The policy document prioritised scientific exploration of the Arctic and the protection of the Arctic's ecological environment. It also underscored two other critical areas of Chinese policy concern in the region. First, the utilisation of Arctic resources – namely, the development of Arctic shipping routes and the establishment of the Polar Silk Road; exploration and exploitation of oil, gas, mineral, and other nonliving resources; the conservation and utilisation of fisheries; and the development of tourism resources. Second, the need for China's active participation in Arctic governance and international cooperation through forums like the AC, as well as multilateral engagements including the Arctic Circle Assembly in Iceland and the Arctic Frontiers in Norway. Throughout the policy document, Chinese enterprises and research institutes were encouraged to participate in Arctic infrastructure development, resource exploration, and collaboration with foreign counterparts.[46]

The document did not include any discussion of the PRC's security interests in the Arctic region. That, however, does not mean that the Chinese are not considering the region from a security and strategic perspective. The publicly available documents of the People's Liberation Army's (PLA) research institutes recognise the Arctic as a strategic commanding height (战略制高点) that overlooks the Northern Hemisphere and which could serve as a potential commercial and military corridor connecting the Pacific and Atlantic Oceans.[47] The Arctic and Antarctica have also been included in Chinese national security discourses through their incorporation into the PRC's national security legislation. Additionally, some members of the Chinese academic community have already proposed the deployment of China's naval forces to the Arctic to gain influence in this geo-strategically important region.[48]

However, the international discourse surrounding Chinese interests in the Arctic seems to be evolving. Some Arctic states are exhibiting resistance to perceived Chinese economic and political influence in the region. A notable example was the strong criticism of Chinese Arctic activities by former US Secretary of State, Mike Pompeo, in 2019. Intelligence agencies in Nordic states have also expressed concerns regarding Chinese presence in the Arctic. Currently, Russia appears to be the only Arctic country where large-scale Chinese projects are successful and face less scrutiny compared to the Western nations. Nevertheless, it is expected that the Arctic will continue to be an integral part of China's foreign affairs in the years to come. General Secretary Xi himself is determined to expand the PRC's role in Arctic affairs by transforming China into a Polar Great Power.[49] This was underscored by the inclusion of references to the Arctic (and Antarctica) in the CCP's 14th Five-Year Development Plan, emphasising the significance of polar scientific research and the ongoing development of the PSR.[50]

Analysis and Sources

When locating the Arctic within the PRC's national strategy of the great rejuvenation, this book aims to accomplish the following: (1) Analyse Chinese Arctic scientific activities and technological capabilities, both domestically and internationally, including the research infrastructure, long-term goals, and the significance of this research for China's understanding of the region, its Arctic identity, and international perceptions; (2) examine the Arctic from the perspective of the PRC's Comprehensive National Security Outlook developed during the Xi era, focusing on its military, economic, technological, and political components; (3) discuss the PRC's official and academic discourses concerning the current Arctic governance system and the views of the region within bilateral relations with Arctic states. The involvement of the PRC's subnational actors, particularly its SOEs and government-linked research institutes, will be included throughout the book, as they represent the most visible presence of the PRC in the Arctic and can serve various functions beneficial to the Party-state. These dimensions broadly represent China's national strategy centred on scientific and technological innovation, the dominant role of security and its linkages to development, and active involvement in international politics. A more detailed discussion of these dimensions will be presented in Chapter 2.

China's engagement with the Arctic region has previously been explored in several book-length analyses. Anne-Marie Brady's influential work published in 2017 examined the PRC's interests in the Arctic and Antarctica, including in the security domain, and its transformation into a Polar Great Power.[51] In an edited volume, Timo Koivurova and Sanna Kopra focused on China's Arctic policy and presence, exploring its scientific, environmental, and economic interests in the region.[52] Nong Hong delved into China's involvement in Arctic governance mechanisms, resource exploration, and shipping routes.[53] These studies collectively highlight the PRC's multifaceted interests in the Arctic and its aspirations for greater involvement in regional governance. This book contributes to the existing literature by providing a nuanced analysis of the Arctic's role in the PRC's national strategy of the great rejuvenation in Xi Jinping's New Era as a science-security-governance nexus as outlined in Chinese documents and as presented in the writings of China's strategic thinkers, scholars, and academics. In doing so, this book adopts an "inside-out" approach in the study of China's global interactions, shedding light on prominent trends, ideas, and views about the Arctic and its importance to the PRC that circulate within the confines of the tightly controlled Party-state system dominated by the CCP.

This book relies on a range of Chinese sources published in Chinese and English. These sources include speeches, major policy addresses, and commentaries by the CCP leadership, PRC officials, and diplomats, as well as white papers issued by the PRC State Council, which serves as the executive branch of the Party-state. It also draws on official Party media, such as the *People's Daily, the PLA Daily, Guangming Daily, Science and Technology Daily and Xinhua News Agency*, used by the CCP to present its views on important domestic and international issues. In addition, the book incorporates what Doshi categorises as "functional sources," including documents released by Chinese ministries, the military, and their associated presses.[54] For instance, Chapter 4 references several editions of the Chinese *Science of Military Strategy* (SMS 战略学) published by the Chinese Academy of Military Sciences and the PLA's National Defence University. While these publications do not contain the PRC's official military strategy, military strategic guidelines, or the PLA's doctrine, they offer insights into the perspectives of leading strategists within the PLA, thereby serving as informative sources on how parts of the Party-state perceive issues related to China's external environment and security.[55] Furthermore, this category includes online releases of official documents, interviews, and news reports from Chinese ministries and ministry-level organisations under the State Council, such as the Ministry of Natural Resources, the Ministry of Foreign Affairs, and the Ministry of Transportation. Zhang et al. argue that these ministries, along with their subordinate units, are at the forefront of China's engagement with the Arctic region. Their websites serve as information hubs that compile records of Chinese Arctic activities, and the meticulous manner in which these releases are posted allows them to reflect official positions on Arctic-related topics to some extent.[56] The analysis presented in this book is further supplemented by primary and secondary English-language sources, including journal articles, books, book chapters, government, and think-tank reports, as well as news outlets.

In addition, the book cites and references a number of "non-authoritative sources," including the works of Chinese scholars, analysts, and academics published in open-source journals and monographs. The inclusion of these sources is important for two reasons. First, given the expanding nature and complexity of China's foreign affairs, Chinese decision-makers from across the Party-state system often require specialised information that they may not possess themselves. In such cases, they can and do turn to researchers and academic institutions for consultations.[57] These scholars are affiliated with numerous specialised research institutions that are directly managed by Chinese ministries and effectively function as their think tanks. For example, the Shanghai Institutes for International Studies (SIIS) falls under the Ministry of Foreign Affairs. There are also several avenues through which researchers and academics from these think tanks or universities can contribute to Chinese foreign policy thinking. These include participation in conferences, attending government meetings, drafting commissioned and non-commissioned (internal – 内部) reports, and being included in senior officials' travel delegations and reports on trips abroad.[58] With regards to the Arctic, for example, when the Chinese MFA needs to address an Arctic-related issue or prepare for an Arctic-specific event, several domestic Arctic scholars are invited to the Ministry in Beijing for brainstorming sessions.[59] It has also been reported that researchers from the SIIS have contributed to the formulation of the PRC's Arctic development plans and official Arctic policy.[60] Furthermore, the writings of the SIIS Vice President, Yang Jian, are often regarded as "the most authoritative unofficial reflections in open-source literature of China's thinking on issues of sovereignty and rights in the region."[61]

Second, while a state's developmental trajectory and its engagement with the international community are shaped by material factors, such as geography, economy, and the policies and actions of other states, it is also important to recognise that what elites within the Party-state think and say about Chinese domestic and foreign affairs can provide valuable insights into the potential developmental paths of the PRC.[62] David Shambaugh, for example, observed that as China gains greater power and prominence on the international stage, "it will concomitantly become more important for foreign analysts to dig deep inside of and understand China's international relations discourse (as well as mass opinion) in order to ascertain China's possible directions and actions."[63] As such, drawing on Daniel Lynch, it is important to analyse the various images – "the discussions and debates concerning

China's possible trajectories" – of the PRC's academic and think tank elites since these "can serve as a mirror on the policy choices the power centre is most likely to face."[64] Even if Chinese elites simply echo the demands of the CCP in their discussions, it is still worth considering these images, especially "how they reflect what is expected and what (possibly in contrast) would be normatively desired" in the assessment of the PRC's trajectories "precisely because the elites are operating inside parameters imposed by the (still) awesomely powerful Party-state."[65] However, it is equally important not to overstate the role and importance of Chinese academic and research elites in the Chinese decision-making process. Ultimately, it is the CCP that holds the authority to determine the fundamental directions that the PRC will pursue.

Organisation of the Book

This book is divided into six chapters. Following this introduction, Chapter 2 outlines Xi Jinping's national strategy of the Great Rejuvenation of the Chinese Nation as a framework for exploring the possible utilities of China's engagement with the Arctic. In particular, the chapter focuses on three dimensions of this national strategy. First, it considers the CCP's efforts to restructure the Chinese economy towards domestically driven consumption based on innovation, high-end manufacturing, and science and technological self-reliance. Second, the chapter outlines Chinese military modernisation, the development of its Comprehensive National Security Outlook, and the strategy of military-civil fusion implemented across the Party-state system. Third, the chapter presents the main ideas and concepts in Chinese foreign policy under Xi, including, for example, the concept of establishing a Community of Shared Future for Mankind. The chapter further highlights Chinese interests in the strategic new frontiers – the deep seas, cyberspace, outer space, and the polar regions – as well as the myriad of bureaucratic actors that exist within the Party-state system to advance Chinese interests in these new frontiers and in the Arctic in particular.

Chapter 3 explores Chinese scientific engagement with the Arctic region and its efforts to establish an integrated Arctic space, atmosphere, land, ice, and underwater multidimensional scientific observation and monitoring network. In doing so, the chapter introduces the most important Chinese domestic scientific and academic institutions engaged in studying the Arctic region, as well as the Chinese Arctic field monitoring infrastructure. This infrastructure comprises various in situ research platforms and remote sensing capabilities, such as stations, icebreakers, unmanned systems, and polar-orbiting satellites. Furthermore, the chapter also discusses Chinese Arctic social science research and some of the challenges faced by the Chinese polar observation program. The remainder of the chapter considers the implications of Chinese Arctic science, including an analysis of the importance of this scientific engagement to the PRC, specifically in terms of advancing China's ecological security and the impacts of Arctic environmental changes, the expansion of its international scientific networks, and contributions to Arctic identity building.

Chapter 4 analyses the PRC's involvement in the Arctic region from the perspective of one of the most important features of China's national strategy – the

Comprehensive National Security Outlook. Specifically, the chapter considers four key types of security from within this outlook. First, it examines military security, delving into how the Chinese armed forces, the PLA, perceive the Arctic and the rationale behind its interests in the region. This section also includes a focus on Chinese strategic science in the Arctic as a tool that can aid the Chinese Party-state in enhancing its understanding and situational awareness of the region. Second, the chapter explores economic security and Chinese efforts to engage in Arctic resource extraction (such as hydrocarbons and minerals) and commercial shipping, primarily through the development of the Polar Silk Road. Third, it discusses technological security and elucidates the Chinese perspective on leveraging the Arctic for testing and developing cutting-edge technologies and manufacturing capabilities. This includes the deployment of unmanned systems and the construction of ice-class ships. Lastly, the chapter addresses political security, which constitutes a fundamental aspect of China's Comprehensive National Security Outlook. It examines how China's engagement in the Arctic can contribute to regime stability by fostering economic development and attaining international status and prestige.

Chapter 5 examines the Chinese official and academic discourses regarding Arctic governance, its challenges, and the normative solutions proposed by Chinese analysts to address them. First, the chapter provides an overview of the current Arctic governance system, which is based on the primacy of Arctic states in regional affairs. It also explores the role and functions of the most prominent regional intergovernmental forum, the Arctic Council, while shedding light on the multitude of challenges confronting the region. Next, the chapter outlines the official position of the PRC on regional governance. It then proceeds by an analysis of Chinese academic discourses regarding Arctic governance. This analysis explores the perceived inefficiencies, challenges, and shortcomings of the current governance system, as well as the solutions proposed by Chinese academics and analysts to enhance Arctic governance going forward. Finally, the chapter examines the PRC's bilateral relations with Arctic states since the extent of China's participation in regional governance is influenced by how these states perceive Chinese activities in the Arctic.

Chapter 6 concludes the book. It provides an overview of the preceding chapters that focused on Chinese Arctic science, security, and governance, as well as the roles of the Chinese state and substate actors in the Arctic. It further reflects on how these are linked to the national strategy of the great rejuvenation and its impact on the PRC's comprehensive national power. Additionally, the chapter discusses the implications of China's deepening interests and presence in the region for its relations with the US and Russia and how this may influence the future development of the broader structures that guide Arctic regional governance.

Notes

1 "中国第8次北极科学考察" [China's Eight Arctic Scientific Expedition], 国家海洋局 [*State Oceanic Administration*], October 13, 2017, www.mnr.gov.cn/dt/hy/201710/t20171013_2333317.html.

2 Yu Qiongyuan, "提升参与国际海洋治理话语权 – 访中国第八次北极科考队领队徐韧" [Enhancing the Right to Speak in International Maritime Governance – Interview with Xu Ren, China's Eight Arctic Scientific Expedition Team Leader], *Xinhua*, July 29, 2017, https://web.archive.org/web/20190717141117/www.xinhuanet.com/2017-07/29/c_1121400008.htm.

3 Thomas Nilsen, "FSB Signs Maritime Security Cooperation with China in Murmansk," *The Barents Observer*, April 25, 2023, https://thebarentsobserver.com/en/security/2023/04/fsb-signs-maritime-security-cooperation-china-murmansk.

4 Xi Jinping, "Secure a Decisive Victory in Building a Moderately Prosperous Society in All Respects and Strive for the Great Success of Socialism with Chinese Characteristics for a New Era," *Xinhua*, October 18, 2017, www.xinhuanet.com/english/download/Xi_Jinping's_report_at_19th_CPC_National_Congress.pdf.

5 "正确理解和大力推进中国式现代化" [Correctly Understand and Vigorously Promote Chinese Style Modernization], 人民日报 [*People's Daily*], February 8, 2023, http://paper.people.com.cn/rmrb/html/2023-02/08/nw.D110000renmrb_20230208_1-01.htm.

6 Vivienne Shue, "Party-State, Nation, Empire: Rethinking the Grammar of Chinese Governance," *Journal of Chinese Governance* 3, no. 3 (2018): 280.

7 Rasmus Gjedssø Bertelsen, Li Xing and Mette Højris Gregersen, "Chinese Arctic Science Diplomacy: An Instrument for Achieving the Chinese Dream?" in *Global Challenges in the Arctic Region: Sovereignty, Environment and Geopolitical Balance*, eds. Elena Conde and Sara Iglesias Sánchez (Oxford: Taylor & Francis, 2016), 444–445.

8 This research adopts the definition of the Arctic region as proposed by the *Arctic Monitoring and Assessment Programme (AMAP)*. The zone covered by the AMAP definition includes the terrestrial and marine areas north of the Arctic Circle (66° 32' N) and north of 62°N in Asia and 60°N in North America, while modified to include the marine areas north of the Aleutian chain, Hudson Bay, and parts of the North Atlantic Ocean, covering an area of approximately 33.4 million km² out of which 60% is defined as Arctic waters. As such, the US, Canada, Denmark, Iceland, Norway, Sweden, Finland, and Russia qualify as Arctic states. See, Arctic Monitoring and Assessment Programme, *AMAP Assessment Report: Arctic Pollution Issues* (Oslo, Norway, 1998), 9. The terms *Arctic, Arctic Region, Circumpolar North*, and *High North* will be used interchangeably to refer to this study area.

9 "Climate Change in the Arctic," *Norwegian Polar Institute*, n.d., www.npolar.no/en/themes/climate-change-in-the-arctic/#toggle-id-6.

10 "Arctic Sea Ice Minimum Ties for Tenth Lowest," *National Snow and Ice Data Centre*, September 22, 2022, http://nsidc.org/arcticseaicenews/2022/09/.

11 Arctic Monitoring and Assessment Programme (AMAP), *Arctic Climate Change Update 2021: Key Trends and Impacts (Summary for Policy-makers)* (Tromsø, Norway: AMAP Secretariat, 2021), www.amap.no/documents/download/6759/inline.

12 On the last point, AMAP notes that "there is currently no agreement in the science community on the degree to which observed changes in the Arctic are directly connected to mid-latitude extreme weather events." See, AMAP, *Arctic Climate Change Update 2021*, 7.

13 "Project Overview – About ArCS II," *Arctic Challenge for Sustainability II*, n.d., www.nipr.ac.jp/arcs2/e/about/.

14 "The Expedition," *MOSAiC*, n.d., https://mosaic-expedition.org/expedition/.

15 For the most widely recognised assessment see: Donald Gautier et al., "Assessment of Undiscovered Oil and Gas in the Arctic," *Science* 324, no. 5931 (2009), www.sciencemag.org/content/324/5931/1175.full.html.

16 The development of Arctic shipping routes is constrained by the harsh Arctic environment, lack of support infrastructure, such as deep-water ports, and emergency response units. For an assessment of the Arctic shipping potential see Rachael Gosnell, "The Potential of Polar Routes: The Opening of a New Ocean," in *Handbook on Geopolitics and Security in the Arctic: The High North Between Cooperation and Confrontation*, ed. Joachim Weber (Cham, Switzerland: Springer, 2020), 193–205.

17 President Putin cited in Gleb Bryanski, "Russia's Putin Says Arctic Trade Route to Rival Suez," *Reuters*, September 22, 2011, www.reuters.com/article/us-russia-arctic-idUSTR E78L5TC20110922.

18 Government of India, *India's Arctic Policy: Building a Partnership for Sustainable Development* (Ministry of Earth Sciences, New Delhi, 2022), www.moes.gov.in/sites/ default/files/2022-03/compressed-SINGLE-PAGE-ENGLISH.pdf.

19 Heather A. Conley, Matthew Melino and Jon B. Alterman, "The Ice Curtain: Russia's Arctic Military Presence," *CSIS*, March 26, 2020, www.csis.org/analysis/ ice-curtain-russias-arctic-military-presence.

20 For an in-depth analysis of the role of conferences in Arctic governance see Beate Steinveg, *Arctic Governance Through Conferencing: Actors, Agendas and Arenas* (Cham, Switzerland: Springer, 2023).

21 For a discussion on China and the Svalbard Treaty see Liu Nengye, "China and One Hundred Years of the Svalbard Treaty: Past, Present and Future," *Marine Policy* 124 (2021): 1–6.

22 "中国人的北极考察活动" [Chinese Arctic Expedition Activities], 国家海洋局极地 考察办公室 [*Chinese Arctic and Antarctic Administration*], December 1, 2018, http:// chinare.mnr.gov.cn/catalog/detail?id=d90c6c2b155b4ec58067cbe92467f876&from=zt fwjdkp&currentIndex=5.

23 Alan K. Henrikson, "The Arctic Peace Projection: From Cold War Fronts to Cooperative Fora," in *Routledge Handbook of Arctic Security*, eds. Gunhild Hoogensen Gjørv, Marc Lanteigne and Horatio Sam-Aggrey (London: Routledge, 2020), 13–25.

24 CAST (中国科学技术协会) clusters Chinese scientific and technological professionals and functions as a bridge that links the CCP with the country's science and technology community. CAST is also a constituent member of the PRC's top political advisory body – the Chinese People's Political Consultative Conference (CPPCC), see "About Us – Profile," *China Association for Science and Technology*, n.d., https://web.archive. org/web/20230319155237/http://english.cast.org.cn/col/col471/index.html#1.

25 Shen Aimin, "北极科学考察筹备进展情况综述" [The Summary of the Progress of Arctic Scientific Expedition Preparations], 学会 [*Learning*] 7 (1994): 17–18.

26 Yan Qide, ed., 走进北极:遥远不是梦 [*Into the Arctic – Not a Dream Anymore*] (Shanghai: 上海科学普及出版社 [Shanghai Popular Science Press], 2009), 160–172.

27 Chen Liqi, "北极在召唤: 中国加入国际北极科学委员会" [The Arctic Is Calling – China Joined the International Arctic Science Committee], 海洋世界 [*Ocean World*] 7 (1996): 22–23.

28 For an overview of the first Arctic scientific expedition see 中国首次北极科学考察队 [China's First Arctic Expedition Team], 中国首次北极科学考察报告 [*CHINARE Report*] (Beijing: 海洋出版社 [Ocean Press], 2004).

29 Gao Dengyi, 梦幻北极 [*Dreaming of the Arctic*] (Shanghai: 上海世纪出版集团 [Shanghai Century Publishing Group], 2005), 74–95.

30 Ma Xiaolan, "我国将在北极建科考站" [China Will Build a Research Station in the Arctic], 科学时报 [*The Science Times*], July 24, 2002, www.cas.cn/ky/kyjz/200207/ t20020724_1026026.shtml.

31 He Jianfeng, "中国北极建站巡礼" [Chinese Arctic Station Construction Tour], 海洋 开发与管理 [*Ocean Development and Management*] 5 (2004): 25–26.

32 Matthew Willis and Duncan Depledge, "How We Learned to Stop Worrying about China's Arctic Ambitions: Understanding China's Admission to the Arctic Council, 2004–2013," in *Handbook of the Politics of the Arctic*, eds. Leif Christian Jensen and Geir Hønneland (Northampton: Edward Elgar Publishing, 2015), 388–407.

33 Alternatively, the PRC's desire to join the AC could have been a result of the CCP's strategy of soft balancing that aimed at China joining various multilateral arrangements in order to limit the ability of the US to impose its preferences on the PRC, see Yuankang Wang, "China's Response to the Unipolar World: The Strategic Logic of Peaceful Development," *Journal of Asian and African Studies* 45, no. 5 (2010): 554–567.

34 "俄插旗推倒多米诺骨牌　各国争夺北极众生相" [Russian Flag-Planting Caused a Domino Effect – Countries Competing for the Arctic], 中国日报网 [*China Daily Online*], August 12, 2007, www.chinadaily.com.cn/hqgj/2007-08/12/content_6022776.htm.

35 Kenneth J. Bird et al., "Circum-Arctic Resource Appraisal: Estimates of Undiscovered Oil and Gas North of the Arctic Circle," *US Geological Survey Fact Sheet 2008–3049*, https://pubs.usgs.gov/fs/2008/3049/.

36 "China's View on Arctic Cooperation," *Ministry of Foreign Affairs of the PRC*, July 30, 2010, https://web.archive.org/web/20150924050143/www.fmprc.gov.cn/mfa_eng/wjb_663304/zzjg_663340/tyfls_665260/tfsxw_665262/t812046.shtml.

37 State Oceanic Administration, 2010年度中国极地考察报告 [*2010 Annual Report on Chinese Polar Research*] (Beijing, China, December 2010), 77.

38 "Agreements and Declarations Signed Following a Meeting Between Prime Minister Jóhanna Sigurðardóttir and Premier Wen Jiabao in Reykjavik Today," *Government of Iceland*, April 20, 2012, www.government.is/news/article/2012/04/20/Agreements-and-declarations-signed-following-a-meeting-between-Prime-Minister-Johanna-Sigurdar-dottir-and-Premier-Wen-Jiabao-in-Reykjavik-today/.

39 "我国第五次北极科学考察队凯旋" [China's Fifth Arctic Scientific Expedition Team Triumphed], *Xinhua*, September 27, 2012, www.gov.cn/jrzg/2012-09/27/content_2234341.htm.

40 Lan Lijun, "Statement by H.E. Ambassador Lan Lijun at the Meeting Between the Swedish Chairmanship of the Arctic Council and Observers," *Arctic Council*, November 6, 2012, https://web.archive.org/web/20191116000746/www.arctic-council.org/images/PDF_attachments/Observer_DMM_2012/ACOBSDMMSE01_Stockholm_2012_Observer_Meeting_Statement_Ambassador_Lan_Lijun_China.pdf.

41 He Miao and Fu Yiming, "中国等6国成为北极理事会正式观察员国" [China and Six Other Countries Becomes a Regular Observer to the Arctic Council], *Xinhua*, May 16, 2013, http://jjckb.xinhuanet.com/2013-05/16/content_445257.htm.

42 Qiu Jiangtao, "中国 – 北欧北极研究中心成立" [China-Nordic Arctic Research Centre Established], 国家海洋局 [*State Oceanic Administration*], December 25, 2013, www.mnr.gov.cn/dt/hy/201312/t20131225_2331697.html.

43 "海洋局副局长出席《中华人民共和国国家海洋局和格陵兰教育、文化、研究和宗教部科学合作谅解备忘录》签字仪式和拜会丹麦外交部北极大使" [The Deputy Director of the State Oceanic Administration Attended the Signing Ceremony of the Memorandum of Understanding on Scientific Cooperation Between the State Oceanic Administration of the PRC and the Greenlandic Ministry of Education, Culture, Research and Church, and also Visited the Arctic Ambassador of the Danish Foreign Ministry], 国家海洋局 [*State Oceanic Administration*], May 16, 2016, www.gov.cn/xinwen/2016-05/16/content_5073689.htm.

44 Zhou Zhiyu, "外交部北极事务特别代表高风：科学外交助力中国参与北极事务" [Special Representative for Arctic Affairs of the Ministry of Foreign Affairs Gao Feng: Science Diplomacy Helps China Participate in Arctic Affairs], 21世纪经济报道 [*21 Century Business Herald*], July 2, 2018, https://web.archive.org/web/20201128012430/http://epaper.21jingji.com/html/2018-02/07/content_80370.htm.

45 "COSCO's M.V. Yong Sheng Completed Her Maiden Voyage Through the Northeast Passage of the Arctic Waters," *COSCO Shipping*, n.d., www.coschartering.co.uk/wp-content/uploads/2015/07/MV-Yong-Sheng-passed-Arctic-Waters.pdf.

46 "Full Text: China's Arctic Policy," *Xinhua*, January 26, 2018, www.xinhuanet.com/english/2018-01/26/c_136926498.htm.

47 Xiao Tianliang, ed., 战略学 [*The Science of Military Strategy*] (Beijing: 国防大学出版社 [National Defence University Press], 2015), 158.

48 For example, see: Yang Zhirong, "北极航道全年开通后世界地缘战略格局的变化研究" [Changes in World Geo-strategic Situation and Countermeasures after the All-Year-Open of Arctic Channel], 国防科技 [*National Defence Science & Technology]* 36, no. 2 (2015): 7–11.

49 Anne-Marie Brady, *China as a Polar Great Power* (Cambridge: Cambridge University Press, 2017).
50 Marc Lanteigne, "The Polar Policies in China's New Five-Year Plan," *The Diplomat*, March 12, 2021, https://thediplomat.com/2021/03/the-polar-policies-in-chinas-new-five-year-plan/.
51 Brady, *China as a Polar Great Power*.
52 Timo Koivurova and Sanna Kopra, eds., *Chinese Policy and Presence in the Arctic* (Leiden and Boston: Brill, 2020).
53 Nong Hong, *China's Role in the Arctic: Observing and Being Observed* (New York: Routledge, 2020).
54 Rush Doshi, *The Long Game: China's Grand Strategy to Displace American Order* (New York: Oxford University Press, 2021), 337–338.
55 M. Taylor Fravel, "China's Changing Approach to Military Strategy: The Science of Military Strategy from 2001 and 2013," in *China's Evolving Military Strategy*, ed. Joe McReynolds (Washington, DC: The Jamestown Foundation, 2017), 45–46.
56 Zhang Qiang, Wan Zheng and Fu Shanshan, "Toward Sustainable Arctic Shipping: Perspectives from China," *Sustainability* 12, no. 21 (2020): 3.
57 Linda Jakobson and Dean Knox, "New Foreign Policy Actors in China," *SIPRI Policy Paper* 26 (2010): 34.
58 Bonnie S. Glaser, "Chinese Foreign Policy Research Institutes and the Practice of Influence," in *China's Foreign Policy: Who Makes It, and How Is It Made?* ed. Gilbert Rozman (New York: Palgrave Macmillan, 2013), 103–113.
59 Martin Kossa, "China's Arctic Engagement: Domestic Actors and Foreign Policy," *Global Change, Peace & Security* 32, no. 1 (2020): 33.
60 Dan Shan, "专访-北极政策白皮书撰写者：中国不觊觎成为北极国家" [Interview with the Writer of the Arctic White Paper: China Does Not Covet to Become an Arctic Nation], 澎湃 [*The Paper*], January 27, 2018, www.thepaper.cn/newsDetail_forward_1970434.
61 Linda Jakobson and Seong-Hyon Lee, *The North East Asian States' Interests in the Arctic and Possible Cooperation with the Kingdom of Denmark* (Stockholm: Stockholm Peace Research Institute, 2013), 8.
62 Daniel C. Lynch, *China's Futures: PRC Elites Debate Economics, Politics, and Foreign Policy* (Stanford, CA: Stanford University Press, 2015), ix–x.
63 David Shambaugh, *China Goes Global: The Partial Power* (Oxford: Oxford University Press, 2013), 14.
64 Lynch, *China's Futures: PRC Elites Debate Economics, Politics, and Foreign Policy*, ix.
65 Ibid., x.

2 The PRC's National Strategy of Great Rejuvenation in the New Era

Socio-Economic, Security, and Foreign Policy Dimensions

Xi Jinping, the General Secretary of the CCP and the core of the Party, has a grand vision for China's development and its position among the great powers of the international system in the 21st century. Since assuming the role of China's most important political figure in late 2012, Xi has tirelessly pursued a national strategy focused on achieving the Great Rejuvenation of the Chinese Nation (中华民族伟大复兴).[1] The concept of national rejuvenation is not unique to Xi and has been a part of Chinese national discourse since the early 20th century.[2] However, as Elizabeth Economy argues, Xi's approach to national rejuvenation differs from that of his predecessors. He has reasserted and strengthened the role of the Party in China's economy and society, centralised authority under his personal leadership, and expanded China's power and role overseas.[3] This has effectively placed the PRC in a new phase of modernisation, development, and internationalisation.[4] The significance of Xi's era at the helm of the PRC was reaffirmed in November 2021 when the CCP passed the *Resolution of the Central Committee of the CCP on the Major Achievements and Historical Experience of the Party over the Past Century*. In this document, which is only the third of its kind in the CCP's 100-year history, Xi's leadership is given equal importance to that of the PRC's most consequential leaders in shaping China's development in the 20th century: Mao Zedong, the founder of the PRC, and Deng Xiaoping, the architect of China's reform and opening policies that led to China's economic and military rise in the early 1990s.[5]

National rejuvenation entails making the PRC a prosperous nation where its citizens thrive, living standards improve, poverty is eradicated, and innovation is the primary driver of development.[6] There is also a persistent emphasis on self-reliance and cultural confidence, allowing China to forge its own path, or go its "own way."[7] On the international stage, Xi aims to utilise Chinese power to shape the international system, align it more closely with the PRC's preferences, and present Chinese ideas and concepts to address global governance issues.[8] Ultimately, China seeks to regain its centrality and prestige in Asia and beyond, as there seems to be an implicit connection between its national rejuvenation and its rise as a global power.[9] Consequently, the PRC's rejuvenation efforts also encompass a strong military component. Xi has undertaken reforms of the Chinese armed forces and advanced their overall modernisation to ensure that the PRC can field a

DOI: 10.4324/9781003295112-2

world-class military by 2049. This is one of the centrepieces of Xi's approach as "a nation cannot be rejuvenated without a strong military."[10]

These ambitions were reiterated and given renewed urgency by Xi during the 19th Chinese Communist Party Congress in October 2017, where he declared that "socialism with Chinese characteristics has crossed the threshold into a new era." According to Xi, this constitutes "a new historic juncture in China's development" propelling the country to the forefront of international affairs. In this New Era, all Chinese citizens are called upon to "strive with one heart to realise the Chinese Dream of national rejuvenation," which will eventually see the PRC as a "global leader in terms of composite national strength and international influence."[11] Han Qingxiang and Chen Shuguan from the Central Party School of the CCP argue that China's rejuvenation will offer the world a model of development that other countries can emulate. It will also introduce a new set of values that challenge the universality of Western values and aim to end [American] hegemony.[12] While the target for achieving national rejuvenation is set for 2049, the Party-state leadership now presents it to domestic and international audiences as an "irreversible historical process."[13]

The purpose of this chapter is to outline the policies and processes that the Chinese Party-state apparatus implements in order to advance the realisation of Xi's grand vision. First, it introduces the CCP's efforts to restructure the Chinese economy towards domestically driven consumption based on innovation, high-end manufacturing, and science and technological self-reliance. Second, the chapter outlines Chinese military modernisation, the development of the Comprehensive National Security Outlook, and the strategy of military-civil fusion implemented across the Party-state. Third, the chapter presents the main ideas and concepts in Chinese foreign policy under Xi including, for example, the concept of establishing a Community of Shared Future for Mankind. Fourth, the chapter analyses the PRC's engagement with the strategic new frontiers as an emerging aspect of China's international affairs. The chapter concludes with an overview of the PRC's foreign affairs bureaucracy and the primary ministerial organisations engaged in China's Arctic activities.

Socio-Economic Dimensions

In terms of domestic economic development, the restructuring of the Chinese economy is a priority for the CCP. This involves focusing on advanced manufacturing, self-reliance, and the development of technologies associated with the Fourth Industrial Revolution, such as artificial intelligence, quantum information, biotechnology, etc.[14] After decades of prioritising high-speed growth through investments in fixed assets (mainly infrastructure) and low-wage/low-end manufacturing for exports, the Party-state leadership is now actively working to rebalance the economy toward high-quality development. This approach aims to be innovation driven to prevent the deceleration of GDP growth.[15] To support these efforts, the CCP has devised and implemented numerous industrial policies and allocated billions of RMB in grants and subsidies to strategic industries. In

2015, the Party-state government launched the *Made in China 2025* action plan, which aimed to promote advanced manufacturing and transform the PRC into a leading manufacturing power. Priority was given to ten sectors, including information technology, ocean engineering and high-end vessels, aerospace equipment, new materials, biomedicine, rail transportation equipment, electric cars, and robotics. The plan placed particular attention on domestic innovation and the development of Chinese brands.[16] According to Economy, this signalled the CCP's early steps toward decoupling the PRC's technological development from foreign suppliers.[17] More recently, in 2020, Xi introduced the *Dual Circulation Strategy* to underscore this reorientation. This strategy aims to insulate the Chinese domestic market from external shocks, eliminate technological bottlenecks, and achieve self-reliance through the country's domestic market.[18] Blanchette and Polk argue that this strategy essentially represents a form of hedged integration, which entails engaging international markets to gain advantages while simultaneously focusing on developing national capabilities to avoid excess reliance on the global economy.[19] Both of these strategies, along with the broader notion of Chinese technological self-reliance, were highlighted in the PRC's 14th Five-Year Development Plan. This plan marks a new, more inward-looking stage of development for the Chinese economy.[20] The construction of the PRC's space station and research icebreaker are presented domestically and abroad as examples of Chinese excellence in advanced manufacturing and the expansion of Chinese high-speed trains, nuclear technology, and 5G networks is considered by the Chinese as evidence of the growing influence of "Made in China" manufacturing in global industrial chains.[21]

The Party-state leadership also recognises the strategic value of data in its economic transformation. The 14th Five-Year Plan for the Development of the Big Data Industry describes data as a national basic strategic resource and a new driving force for economic development. Going beyond national considerations, the plan also includes a section on standardisation that promotes exchanges and cooperation between big data organisations in the PRC and abroad. It encourages Chinese enterprises, universities, research institutes, and industry organisations to actively participate in the formulation of international standards for big data.[22] In fact, the Chinese government has expressed interest in the internationalisation of Chinese technological standards, primarily through the development of the *China Standards 2035* project.[23] This ambition is reinforced by the PRC's increasing technological capabilities and the success of Chinese tech companies on the global stage, exemplified by Huawei's dominance in 5G technologies. Arjun Gargeyas, a research analyst at the Takshashila Institution in India, explains that while *Made in China 2025* aims to solidify China's leading position in global manufacturing and supply chains of advanced technologies, the Standards 2035 project seeks to establish and control the governing framework for the utilisation of such technologies.[24] This would provide the Chinese Party-state with several economic and political advantages. Chinese-developed international standards would facilitate the further global expansion of Chinese tech companies by granting them easier access to international markets. Moreover, these standards could assist in advancing the creation of Chinese-sponsored technological blocs, particularly in the developing world, through cross-border agreements.[25]

The leadership of the CCP acknowledges the critical role of scientific research and domestic technological advancements in realising the vision of national rejuvenation and attaining global influence. General Secretary Xi emphasised this understanding by stating that

> to become strong and rejuvenated, China must vigorously develop science and technology and strive to become the world's major scientific centre and innovation hub. We are closer than ever before to the goal of the Great Rejuvenation of the Chinese Nation, and more than ever before, we need to build ourselves into a world power in science and technology.[26]

For the Chinese, innovation derived from science and technological advancements is crucial to achieving their objective of high-quality development. This includes the development of emerging industries, the upgrading of traditional industries, the construction of major transportation and industrial projects, such as the high-speed railway network, the enhancement of national enterprise competitiveness, and the promotion of regional innovative development.[27] Science and technology innovation also plays a crucial role in the modernisation and expansion efforts of the PLA. Chinese research institutes cooperate with the PLA on technologies that improve the Party-state's military capabilities, particularly in ocean situational awareness and underwater warfare.[28] Simultaneously, China recognises that in today's world, scientific and technological innovation has become one of the primary frontlines of international great power competition and that such competition in areas related to core technologies of the modern industrial age is unprecedented.[29] As argued by a commentary in the *People's Daily*, the only path forward for the PRC in this competition is to achieve self-reliance in science and technology. This is because self-reliance is the key to making the PRC strong and prosperous, and ultimately enabling the Chinese nation to prevail in the international struggle.[30]

Chinese elite debates on the rise and fall of great powers also encompass discussion on the role of science and technology in those processes. This is because industrial revolutions throughout history have impacted the distribution of world power, with a state's science and technology strength determining changes in global political and economic dynamics.[31] For instance, the Chinese argue that the US capitalised on the Third Industrial Revolution (the Digital Revolution) to increase its power and extend its hegemony.[32] Now, some in China see the Fourth Industrial Revolution as an opportunity for the PRC to surpass Western powers and establish global technological leadership.[33] In pursuit of this goal, the CCP plans to transform the country into a global leader in science and technological innovation,[34] allocating billions of RMB to China's research and development (R&D).[35] In 2021, China allocated a record-breaking 2.44% of its GDP to R&D, putting it on track to outspend its only remaining competitor in this field, the US, within this decade.[36] The areas receiving the most funding are those closely associated with the Fourth Industrial Revolution, including space exploration, nuclear physics, quantum science, artificial intelligence, and biological engineering. If the PRC were to become a global leader in these areas, it would gain a significant competitive advantage over its rivals and be able to set technological standards, resulting in market dominance and substantial returns through royalties.[37]

A crucial aspect of the CCP's drive for national rejuvenation is the cultivation of a large and high-quality pool of technologically skilled talent. The *People's Daily* notes that history has demonstrated that possessing first-class innovative talents and scientists confers an advantage in scientific and technological innovation.[38] However, a recent report by the Institute of International and Strategic Studies at Peking University, on the strategic technological competition between the PRC and the US, concluded that while China's technological capabilities are steadily increasing, the US still outperforms China in talent acquisition.[39] To attract and develop talent in the PRC, the Party-state leadership has implemented numerous talent programs, such as the Thousand Talents Plan,[40] aimed at fostering a high-quality workforce and accelerating China's transformation into a major global hub of professional talent.[41]

However, the Chinese science system also exhibits strong connections between science, nationalism, patriotism, ideology, and the CCP.[42] General Secretary Xi posits that in the New Era, Chinese scientists should uphold the spirit of patriotism, respond to the Party's call, and be loyal to the motherland.[43] This perspective sometimes places Chinese scientists at odds with the curiosity-driven and open science system favoured in the US and the EU.[44] Scientific research in the PRC operates within the boundaries of the Party's ideological guidance, with Chinese scientists often required to dedicate their time to Party education and Party-building efforts.[45] Consequently, the Party wields significant influence over all aspects of the PRC's science and technology ecosystem.[46] Chinese knowledge production is ultimately driven by the Party-state's pursuit of national rejuvenation,[47] leading to a utilitarian view of science that regards it as a tool in service of CCP-defined economic and security needs.[48] This perspective is evident in the Party's belief that national laboratories, universities, and leading enterprises in science and technology are essential components of China's strategic scientific and technological forces.[49]

Security Dimensions

In the realm of security, one of the most notable features of Xi's tenure as the core of the CCP and the leader of the PRC is his unwavering focus on national security, considering it "a matter of prime importance."[50] To safeguard the PRC's national security, the CCP continues its efforts to modernise Chinese armed forces across all domains. Xi's aspiration is for the PLA to attain the status of a world-class military capable of projecting power globally, across various regions and continents.[51] Under his guidance, the PLA underwent organisational reforms aimed at enhancing operational effectiveness and streamlining the chain of command. This involved dismantling the long-standing system of general departments, military area commands, and force composition focused on land forces,[52] but also resulted in tighter Party control over the PLA.[53] Furthermore, the PLA bolstered its power-projection capabilities through the introduction of aircraft carriers, long-range bombers, overseas military bases (such as Djibouti),[54] and advanced military technologies, including, but not limited to, fifth-generation stealth fighter jets and hypersonic gliding vehicles.[55] The PRC is also expanding its nuclear capabilities, with a rapidly expanding silo-based intercontinental ballistic missile force, enhanced survivability of its sea-based deterrent (maintaining six Jin-class ballistic missile submarines), and potentially having the capacity to deploy over 700 nuclear warheads by 2027.[56] Chinese armed forces actively engage in military diplomacy, participating in exercises in Southeast Asia, the Indian Ocean, Africa, the Mediterranean, and the North Atlantic, among other locations.[57] These activities and growing capabilities are financed by consistent and continuous growth in the Chinese military budget. According to the Stockholm International Peace Research Institute (SIPRI), the PRC is the world's second-largest spender, allocating an estimated USD 293 billion to its military in 2021, reflecting a 72% increase since Xi Jinping assumed the role of CCP General Secretary in 2012.[58]

The Party-state leadership has also expressed its desire for the PLA to play a more prominent role on the international stage, advancing foreign policy objectives, and safeguarding overseas interests, including Chinese citizens, investments, trade routes, and energy sources.[59] This requires a sustained focus on the maritime domain due to the PRC's heavy dependence on global oceans. Approximately 90% of the PRC's foreign trade and the import of commodities, such as iron ore and coal, are conducted via sea routes.[60] Consequently, there are increasing demands for the Chinese Navy to operate at greater distances from the Chinese mainland. In 2019, the US Indo-Pacific Command estimated that in the preceding two and half years, the Chinese Navy had conducted more global naval deployments than in the previous 30 years.[61] According to foreign analysts, the PLA Navy's objectives now include establishing control over the PRC's near-seas region (including Taiwan), regulating foreign military activity in its Exclusive Economic Zone (EEZ), particularly in the South China Sea, defending the PRC's commercial sea lanes of communication (SLOCs), countering US influence in the Western Pacific, and asserting the PRC's status as a major world power.[62] To achieve these objectives, the CCP has allocated billions of RMB to the PLA's naval modernisation efforts, encompassing a wide array of vessels, aircraft, and weapon systems, as well as improvements in logistics, doctrine, personnel education, training, and exercises.[63] The PLA currently operates the world's largest navy, comprising 350 ships and submarines, the world's largest coast guard, and the leading maritime militia.[64] To meet the construction demands of such a fleet, the Party-state is expanding the naval shipbuilding capacities of existing shipyards, such as Jiangnan in Shanghai.[65] The Chinese Navy is well on its way to becoming a true blue-water navy, possessing the capability to challenge the US Navy on the open seas. The PLA Navy's global ambitions are also apparent in the extensive scope of its ocean survey activities. The PRC operates a vast ocean survey fleet that collects valuable information on the maritime environment in major seas worldwide, including the Pacific, Indian, and Arctic Oceans. This data can be used to enhance the Chinese Navy's situational awareness.[66] Additionally, the Chinese Navy can rely on a global

network of Chinese-owned and operated international port terminals that can provide, in peacetime, dual-use capabilities to the PLAN in the form of logistics and
intelligence hubs.[67] An article in the *PLA Daily*, the official newspaper of the PLA,
boldly asserts that "without powerful overseas military projection capabilities, the
protection of overseas trade and interests would become mere scraps of paper."[68]

This sustained focus on the maritime domain also aligns with Xi's vision of
transforming the PRC into a maritime great power. However, in the Chinese context, being a maritime great power entails not only possessing a formidable military
but also commanding a substantial and effective coast guard, operating a global
merchant marine, possessing significant shipbuilding capacity, and having the ability to exploit vital maritime resources.[69] Li Jian from the PLA Naval Research
Institute argues that making China a maritime great power is a core objective, as
countries that develop maritime capabilities and naval power can more effectively
safeguard their national security and expand to become great powers. Those that
fail to do so decline.[70] Yoshihara and Holmes observe that this understanding of
sea power in a state's national strategy resembles the ideas espoused by the 19th-
century naval strategist Alfred Thayer Mahan, who studied the development of
maritime powers.[71]

However, since assuming the role of General Secretary of the CCP, Xi Jinping
has broadened the concept of national security beyond military affairs to encompass almost all other policy domains. This expansion is reflected in the Comprehensive National Security Outlook (总体国家安全观) adopted in 2014.[72] This
outlook encompasses 16 types of security, including political, territorial, military,
cyber, cultural, economic, societal, technological, ecological, overseas interests,
resource, nuclear, deep-sea, space, polar, and biosecurity.[73] Among these, political security is considered the most fundamental, as it relates to the security of the
regime itself, namely, maintaining the leadership and ruling status of the CCP and
the socialist system with Chinese characteristics in the PRC. According to the CCP,
this is the best way to protect national interests.[74] The adoption of the Comprehensive National Security Outlook aimed to integrate security and developmental
policy objectives.[75] Xi previously stated that "development is the foundation of
security" and "security is the condition for development."[76] Yuan Peng, President
of the influential China Institutes of Contemporary International Relations, argues
that security and development are intertwined and must be planned in tandem to
ensure the stability of the PRC's development.[77] The underlying logic of the Comprehensive National Security Outlook is to address the increasing number of domestic and international risks, including political, social, economic, military, and
ideological factors that could potentially hinder the PRC's modernisation process
and, consequently, national security, as the state's development space and interests
have expanded over the past decade.[78] Simultaneously, the adoption of this security
outlook means that anything perceived by the CCP as an obstacle to achieving its
objectives can be deemed as a national security threat, warranting the involvement
of the Chinese armed forces, both domestically and internationally.[79]

At the intersection of science, security, and development in the PRC lies the
CCP's strategy of Military-Civil Fusion (MCF 军民融合),[80] which was elevated

to a national strategy by Xi in 2014 and integrated into the overall development of the Party and the state. This strategy aims to achieve several objectives, including the joint construction, use, and sharing of military and civilian infrastructure and resources, the fusion of defence-related and civilian industrial sectors, the integration of military and civilian science and technology innovation bases, a two-way training, exchange, and use of military and civilian talents, as well as the utilisation of civilian logistical capabilities for military purposes.[81] The PRC's 2015 Defence White Paper further clarifies that MCF promotes "joint exploration of the sea, outer space, and air, and shared use of such resources as surveying and mapping, navigation, meteorology, and frequency spectra."[82] The CCP's 14th Five-Year Development Plan also emphasises the need to strengthen military-civilian collaboration and innovation in science and technology, particularly in emerging areas, including the maritime domain, space, and technologies of the Fourth Industrial Revolution.[83]

As a national-level strategy, MCF recognises the value of science and technology as the foundation of an advanced military and society and implies a two-way exchange of knowledge and technologies between the military and civilian sectors.[84] It aims to achieve better synergy between national security and economic development in a complex international environment characterised by great power competition.[85] Chinese defence analysts believe that under conditions of intense great power competition and changes in the global power structure, the state that will ultimately prevail is the one with a national governance system that is more adaptable and able to achieve a better integration of its economic and military systems.[86] They argue that for the PRC, MCF represents the only choice to balance national security and economic development in order to "take the initiative and win the future."[87] The sharing of military and civilian resources at major Chinese shipyards, such as Dalian or Jiangnan, where naval vessels, including aircraft carriers, are constructed right next to commercial container ships, provides a good example of MCF in action.[88]

Other great powers, including the US, have similar civilian-military integration plans in place. However, according to Mulvaney, the PRC's MCF is "far deeper and more complex." It is "a state-led, state-directed program and plan to leverage all levers of state and commercial power to strengthen and support the armed wing of the Communist Party of China," the PLA.[89] The Chinese MCF system comprises organisational units from all levels of the Party-state, including State Council ministries, such as the Ministry of Industry and Information Technology, the Central Military Commission's Equipment Development Department, provincial and local governments, such as Hebei, Jiangsu, and Sichuan, state-owned enterprises, such as the Aviation Industry Corporation of China (AVIC),[90] as well as science, research, and educational institutions, such as the Harbin Engineering University.[91]

Foreign Policy Dimensions

In terms of foreign policy, the CCP is endeavouring to shape its external environment to be beneficial to Xi's stated goal of national rejuvenation. This would involve the PRC regaining power, respect, prestige, and influence over Asian politics, and by extension international politics. As mentioned earlier, while Xi is not the first Chinese leader with such goals, it is his approach that sets him apart. Xi's PRC is profoundly more active in international affairs, whether related to development, security, climate, or other areas. It is also much more forceful in safeguarding what it perceives as its core interests – state sovereignty, national security, territorial integrity, national reunification, the preservation of the PRC's political system led by the CCP, economic development, and social stability.[92] Many Chinese experts believe that this change in the PRC's behaviour began after 2008. Ling Wei, the Director at the Zhou Enlai Diplomatic Studies Centre of the China Foreign Affairs University, argues that the overall situation in which Party-state leaders make decisions underwent "subtle but significant changes." The impact of the global financial crisis severely affected the so-called West, while the PRC's material power and international prestige significantly improved, prompting the Party-state leadership to declare its intent to be more proactive in global affairs.[93] This trend continued and was amplified under Xi. In her study of the PRC's foreign policy under Xi, Sørensen notes that Chinese diplomats began to use language that reflected this proactive approach, such as 积极外交 "active diplomacy," 主动进取 "take the initiative," or 更加积极 "be more active."[94] It became apparent that the foreign policy strategy of "keeping a low profile," devised by Deng Xiaoping in the early 1990s, was discarded, and a new one, "striving for achievement," was being put in place. According to Yan Xuetong, this new strategy is cooperative in nature but assertive in practice as it seeks to enlarge the circle of states that are more receptive to the PRC's foreign policy initiatives and actively shape the external environment to the CCP's preference.[95] Chinese analysts argue that in the New Era, as the PRC becomes more confident and assumes a more central role on the world stage, its shaping power will grow, transforming the country into a shaper of a new type of globalisation, where the PRC will be able to formulate new international rules and reform global governance.[96]

The notion that the current global governance system, led by the West, is in dire need of reform has become a recurring theme in Chinese commentaries on international affairs.[97] The CCP leadership maintains that the world is undergoing great changes unseen in a century (百年未有之大变局).[98] Zhou Li, a high-ranking CCP official and a former Vice Minister of the International Department of the CCP, notes that the underlying driver of these changes is the rebalancing of the international system from the West to the East, exemplified in the idea of a rising East and a declining West (东升西降). According to Zhou, the global economic map is being redrawn due to the rise of emerging markets, led by the PRC. The economic centre of gravity is shifting from the Euro-Atlantic region to Asia, the influence of liberal democracy, as represented by the West, is waning, and developing nations are seeking a greater say in global affairs.[99] Additionally, Xi points out that the world is facing new challenges, such as pandemics, new security issues like the war in Ukraine, a weak global economic recovery, a widening developmental gap, and governance deficits in issues such as climate change.[100] Chen Zhimin, the Vice President of Fudan University, argues that these issues require multilateral cooperation, which the current global governance system, grounded in what he calls the

West's selective multilateralism, is unable to address.[101] However, the CCP leadership, perceiving global governance as "unjust and improper,"[102] has a clear vision of how to reform it and address common challenges.

The Party-state diplomatic apparatus actively promotes new ideas and thinking in international affairs. This is reflected in the provision of Chinese-specific solutions and wisdom to steer the reform of global governance. For example, over the past decade, the PRC has expressed its desire to establish a new type of international relations (新型国际关系). This new type would be based on mutual respect as the premise, fairness and justice as the norm, and win-win cooperation as the goal. Chinese analysts argue that this approach transcends the narrow focus on state interests represented by the Western Realist school of international relations.[103] The development of these normative concepts aims to differentiate the PRC from the experiences of the so-called West, which is portrayed as having a zero-sum understanding of world politics, with developed states taking advantage of developing nations. In contrast, as argued by Zhao Kejin, the Deputy Dean of the School of Social Sciences at Tsinghua University, the PRC, as its power grows, will not seek hegemony but instead pave a new great power path of common development for its people and the people of the world.[104] The message is clear: the West represents the old way of thinking about international affairs, grounded in a Cold War–era mentality of ideological blocs and confrontation, while the PRC's rise and development are seen not as a threat but as "an opportunity for the world."[105]

At the practical level, these solutions and wisdom are reflected in various initiatives and platforms launched by the PRC under Xi, including the BRI, the Asian Infrastructure Investment Bank (AIIB), the Global Development Initiative (GDI), the Global Initiative on Data Security, and the Global Security Initiative (GSI). Among these, the BRI stands out as one of the most important projects of the Chinese Party-state leadership, extending beyond its foreign policy apparatus. Launched by Xi in 2013 through a series of speeches in Kazakhstan and Indonesia, the BRI focuses on international connectivity through massive investments in infrastructure, greater intergovernmental policy coordination, the removal of any obstructions to trade and the flow of goods, the deepening of financial integration, and strong people-to-people exchanges.[106] The Initiative spans a vast geographical area, covering much of the Eurasian continent, Africa, Latin America, and Oceania, as well as the Indian and Arctic Oceans. As of February 2022, 149 countries and 32 international organisations signed onto the Initiative in some form.[107] However, there is also a strategic perspective to the BRI. As the Initiative encompasses not only loans and grants for infrastructure development but also involves free-trade negotiations, common industrial standards, energy agreements, student scholarships, joint research centres, currency-swap agreements, and professional training centres, Rolland argues that the CCP aims to establish deep bonds with participating states and draw them into its sphere of influence.[108] Chinese analysts further emphasise that the BRI will enable the PRC to break through Western-imposed norms on trade and investment while simultaneously enhancing the competitiveness of Chinese companies and promoting the development of China's poorer western regions.[109]

Going beyond providing solutions to perceived developmental and economic gaps, General Secretary Xi proposed the establishment of the GSI in 2022 to address the growing global security governance deficit. According to Wang Yi, the PRC's Minister of Foreign Affairs, the GSI adheres to the principles of common, comprehensive, cooperative, and sustainable security. It proposes, among other things, sovereignty as the principal condition of peace, the resolution of conflicts through dialogue, and insists on the indivisibility of security. This means that the security of one country cannot come at the expense of the security of other countries, and regional security cannot be guaranteed by strengthening or even expanding military blocs.[110] The GSI serves as another example of the CCP's efforts to infuse international affairs with what it perceives as unique Chinese ideas and supposedly new thinking for solving global governance issues. An article in the *People's Daily* argues that the GSI demonstrates Chinese wisdom in safeguarding peace because it is steeped in the profound Chinese culture of harmony and incorporates holistic thinking and dialectical principles of harmony to solve the global security dilemma.[111]

Another avenue through which the PRC can advance some of its foreign policy objectives and provide Chinese solutions and wisdom is international organisations. Zhang Guihong, the Director of the Research Centre for the United Nations and International Organisations at Fudan University, suggests that the PRC should actively utilise organisations at both the international and regional levels to promote its initiatives, organise conferences, and recommend Chinese nationals for staff positions.[112] In recent years, Chinese nationals have indeed held leadership positions in various organisations within the United Nations (UN) system, including the International Telecommunication Union and the Food and Agriculture Organisation. References to Chinese proposals, such as the BRI, have been inserted into the UN's Sustainable Development Goals and have appeared in UN resolutions, indicating the level of influence exerted by the PRC within the UN.[113] Beyond the UN system, the PRC actively engages within the BRICS group – an association of emerging economies consisting of Brazil, Russia, India, China, and South Africa – and aims to leverage it as an additional platform to promote its vision of the world. During the 2022 summit, when addressing the leaders of the BRICS nations, Xi stated that the PRC would like to collaborate with its BRICS partners to implement its Global Security and Development Initiatives, aiming to "inject stability and positive energy into the world."[114]

In terms of bilateral partnerships, the PRC's relations with Russia play a crucial role in implementing many Chinese initiatives. Both countries share a similar understanding of international security, development, human rights, and democracy as there are "no forbidden areas of cooperation" between these two powers.[115] Russia has already expressed support for the initiatives proposed by the PRC, including the GDI and the GSI. Their partnership involves cooperation and coordination across various areas, with a particular interest in their mutual engagement in military affairs, despite historical mistrust. Their armed forces regularly engage in joint training, and the two powers recently reached an agreement for Russia to assist the PRC in developing a missile launch detection system.[116] Foreign observers note that the PRC leadership views Russia as an asset that can be leveraged in China's long-term, ideologically driven competition with the US and Western liberalism. As such, the CCP leadership is unwilling to dial back its political alignment with Russia, even after Russia's invasion of Ukraine in early 2022.[117]

What unites all of these initiatives is the belief that humanity needs to come together under the (Chinese) banner of the Community of Shared Future for Mankind (人类命运共同体).[118] The Party-state leadership has been vocal in recent years in criticising the current global governance system, particularly the role of the West in it, while expressing more confidence in the PRC's strength and ability to provide competing models of development and governance.[119] Zhao Xiaochun, a Professor at the University of International Relations in Beijing, argues that given the PRC's changed perception of its international status and the West's governance failures, it is imperative to seek alternative models of governance that are more just and representative.[120] The Community of Shared Future for Mankind is the Chinese solution to the world's current challenges. Originally mentioned by General Secretary Xi during his visit to Russia in 2013, the concept was further developed in Xi's speech at the UN office in 2017,[121] coinciding with the PRC's entry into its New Era. The Community is envisioned to be based on mutual respect, win-win cooperation, consultations, and dialogue as mechanisms for resolving disputes. It also promotes inclusive economic globalisation and partnerships rather than alliances and opposes Cold War mentality and power politics.[122] According to Yang Jiechi, the PRC's former Minister for Foreign Affairs and former member of the Politburo of the CCP, this reflects the PRC's desire to contribute even more to human progress as China approaches the centre of the world stage.[123] Simultaneously, the construction of this Community aligns with the PRC's efforts for national rejuvenation, creating a favourable external environment for the country. While the PRC leadership claims that within the Community all countries, regardless of their social systems, histories, and ideologies, would align their goals and interests and share international responsibilities for humanity's progress, it is evident that the concept also incorporates the CCP's security and developmental needs. According to Rolland, the reference to partnerships rather than alliances in the political sphere is a criticism of the US alliance system perceived as a threat to the PRC's survival. References to Cold War mentality are aimed at countering outside efforts to contain or undermine the PRC's rise, while the emphasis on win-win cooperation and an inclusive global economy reflects the PRC's desire to maintain unrestricted access to overseas markets and energy sources.[124] It is through the process of building a Community of Shared Future for Mankind that the PRC aims to establish an alternative governance model capable of attracting numerous countries, primarily from the global South, while also providing the impetus for China to shape the international order.

The PRC's efforts to shape its external environment are accompanied by the active development of its discourse power, commensurate with China's increased comprehensive national strength and international status. Discourse power, a concept rooted in a state's material power, reflects the PRC's aspiration to have the right to speak on matters related to international affairs and inject Chinese ideas, concepts, and propositions (such as the Community of Shared Future for Mankind) into international narratives. This is to influence international perceptions of China and ultimately to shape the norms that underpin the global governance system.[125] The CCP leadership places significant importance on China's international discourse power. It believes it necessary to elucidate the PRC's vision of development, security, human rights, ecology, and global governance, and "illustrate to the world that China's development in itself is its greatest contribution to the world."[126] Chinese analysts argue that the current global governance system remains dominated by Western discourses (on human rights or socio-economic development for example), which have allowed the developed world to formulate international rules and exert control over much of the non-Western world.[127] The PRC now seeks to challenge this perceived Western discourse hegemony by offering its own solutions to various global governance deficits and effecting changes in the international order.[128] He Yiting, the former Vice Dean of the Central Party School, contends that the rise of a state's material capabilities must be accompanied by a rise in its discourse power. Otherwise, that country cannot be considered a modern great power.[129] Consequently, the PRC will advocate for the need for the international community to adopt the solutions and wisdom provided by its Community of Shared Future for Mankind, marginalising what it perceives as Western discourses and reforming what it considers an unbalanced global governance system.

To enhance the effectiveness of its discourse power, the CCP is vigorously developing distinct Chinese social science practices and theoretical innovations – rooted in China's unique history, culture, and socialist developmental practices and experiences – aiming to increase the PRC's worldwide appeal in terms of soft power.[130] The importance of developing distinct Chinese social sciences was emphasised during Xi's visit to Renmin University in April 2022. During the visit, the General Secretary urged Chinese universities to avoid replicating foreign standards and models, instead focussing on "developing philosophy and social sciences with Chinese characteristics" to establish "an independent knowledge system."[131] According to Xie Fuzhan, a researcher at the Chinese Academy of Social Sciences, such calls indicate that under CCP leadership, the PRC is confident in its ability to develop Chinese solutions and offer them to countries worldwide.[132]

While the CCP portrays national rejuvenation as an irreversible process and changes in the global balance of power as the prevailing trend, the PRC's transition into a global superpower is encountering challenges. Domestically, the CCP must grapple with a demographic decline and an aging population, which impacts productivity and places additional burdens on the state's social and healthcare systems.[133] The growth of the PRC's economy is decelerating, and the double-digit GDP figures witnessed over the past three decades are not expected to return. The PRC has entered a "new normal" phase of slower growth, focussing on quality rather than quantity.[134] More recently, the PRC has experienced supply shocks and port closures due to lockdowns enforced by the CCP as part of its zero-COVID policy to curb the spread of the COVID-19 virus, further negatively affecting its economic outlook.[135] Internationally, the PRC is engaged in a great power competition with the US, and international perceptions of the CCP-led country are increasingly negative, particularly in the so-called developed world.[136] Formal and informal groupings of like-minded states have emerged in the past decade to address an assertive PRC, including AUKUS – a defence cooperation arrangement between the US, UK, and Australia – and the Quad – a partnership between the US, Japan, Australia, and India. The cooperation and coordination among these states extend beyond defence-related matters and increasingly encompass areas such as technological standard settings.[137] Consequently, Chinese state-linked companies like Huawei or ZTE face heightened scrutiny, and there is potential for their exclusion from the development of communication networks in Europe or North America.[138]

Nevertheless, the CCP is confident regarding the future prospects of the PRC, its political system, rising international influence, and position vis-à-vis the US.[139] This confidence stems from the belief of the Party-state leadership that the PRC possesses five strategic advantages for its development: (1) strong leadership of the CCP, (2) a socialist system with Chinese characteristics, (3) a solid foundation for sustained and rapid development (a large economy), (4) a stable social environment, and (5) a spiritual strength based on self-confidence and self-improvement.[140] Additionally, Ma Jiantang, the Party Secretary of the Development Research Centre of the State Council, and Zhao Changwen, the Director of the Centre for International Knowledge on Development, argue that under CCP leadership, the PRC has established a new form of human civilisation that broke the centrality of the West while avoiding its shortcomings, such as political polarisation or partisanship.[141] These perceived institutional advantages enable the PRC to pursue its national rejuvenation, extend the period of strategic opportunity for its development, and ultimately succeed in the competition among nations.[142] Therefore, even though the world around the PRC enters a period of turbulence, disorder, and uncertainty,[143] General Secretary Xi is confident that time and momentum are on China's side[144] and that opportunities outweigh challenges.[145]

Strategic New Frontiers in the Xi Jinping Era

The PRC, under the leadership of General Secretary Xi, has attached significant political, economic, and military importance to the strategic new frontiers – cyberspace, the polar regions, the deep sea, and outer space. In 2015, the Chinese government recognised that China had substantial security and economic interests in these new frontiers. Simultaneously, the Party-state acknowledged that the country faced security threats and challenges within these frontiers, necessitating the establishment of mechanisms to safeguard its national security interests.[146] The subsequent National Security Law (2015, Article 32) called for the Party-state to pursue peaceful exploration and utilisation of these domains while enhancing capabilities to safeguard the security of its operational assets, ensure safe access, promote scientific exploration, and exploitation of the new frontiers.[147] Since then, references to the strategic new frontiers, their importance, and the need for active PRC participation in their governance have been included in speeches and documents produced by high-level CCP officials related to global governance.

Within the broader Chinese elite discourse, there are several depictions of the strategic new frontiers. Chinese scholars primarily view them as areas of great power competition for resources and international influence.[148] Zhang Zhijun and Liu Huirong from the Ocean University of China argue that the strategic new frontiers are important battlefields in future global power competition, and their strategic significance should be recognised at the national level and included in overall national planning.[149] The authoritative 2020 edition of the *Science of Military Strategy* emphasises that the strategic new frontiers "are not only important domains of military conflict but also integral parts of international political struggle, significantly influencing a country's politics and diplomacy. Therefore, all countries, particularly the great powers, closely monitor and plan military conflicts in these new areas."[150] Chinese analysts also stress the connections between these domains and perceived US hegemony. For instance, Carla P. Freeman, in her analysis of Chinese positions on maritime and outer space domains, highlights that Chinese scholars believe these domains serve as enablers of US global dominance.[151] Similarly, Deng Beixi from the Polar Research Institute of China argues that the establishment of marine protected areas in the Southern Ocean surrounding the Antarctic continent, many proposed by the US, effectively acts as a form of "soft control" dividing the southern polar region into spheres of influence and restricting economic and scientific activities of other nations, including the PRC.[152]

Beyond purely military and great power competition connotations, the strategic new frontiers are also seen as resource-rich domains capable of supporting a state's economic growth and sustainable development.[153] Shi Xianpeng and Wu Changbin from the National Deep Sea Centre – an organisation under the PRC's Ministry of Natural Resources – assert that the deep sea holds a treasure trove of resources and potential as a strategic base for ensuring sustainable social and economic development, conducting scientific and technological innovation, and safeguarding national security.[154] The PRC's white paper on China's space program further argues that engagement with outer space and subsequent development of space science and technology will directly contribute to industrial production and boost China's overall growth.[155]

Finally, Chinese analysts criticise the current governance mechanisms in the strategic new frontiers as inefficient and in need of new ideas and solutions to upgrade the governance structures. Yang Jian from the Shanghai Institutes of International Studies argues that the West, particularly the US, failed to provide the ethical basis and corresponding public goods for effective governance in the new frontiers. Instead, the West pursued a hegemonic approach and expanded its power space through wars, conquest, colonisation, and competition. For example, the modern Western expansion in the maritime domain was a result of colonisation conducted in the name of discovery and exploration. The author contends that such a Western hegemonic approach persists today in the new frontiers. Yang asserts that the US promotes unilateralist security thinking, leading to a security dilemma between major powers in the new frontiers. This hegemonic thinking and unilateral understanding of security hamper the development of common security in the new frontiers, resulting in confusion and an entire governance process in the new frontiers lacking ethics and morality.[156] Chinese academics also highlight the potential impacts of the militarisation of the new frontiers on the larger international community. Dong Yongzai, a researcher at the Chinese Academy of Military Sciences, notes that the military power play between the US and Russia in the polar regions poses a direct challenge to the principle of polar demilitarisation and, more importantly, endangers the peaceful use of the polar regions and territorial defence of all countries, including the PRC.[157]

Patrik Andersson from Aalborg University in Denmark, in his analysis of the Chinese discourse on the strategic new frontiers concept, points out that the label "strategic" in strategic new frontiers denotes a particular form of importance – an understanding that the new frontiers are important from the perspective of larger strategic considerations and that they are connected to the PRC's national security.[158] The strategic connotation of the new frontiers has become clear in recent years as all four domains – cyberspace, the polar regions, the deep sea, and outer space – have been connected to the PRC's BRI. Under the umbrella of the BRI, China is now constructing a Digital Silk Road (数字丝绸之路) in cyberspace, a Polar Silk Road (冰上丝绸之路) in the Arctic, the deep sea development can be linked to the Maritime Silk Road (海上丝绸之路), and regarding outer space, a BRI Space Information Corridor (一带一路空间信息走廊) is being formed. Additionally, the PRC aspires to become a great power with international influence in all these frontiers. The CCP leadership has repeatedly expressed its intention to transform the PRC into a cyber great power (网络强国),[159] polar great power (极地强国),[160] maritime great power (海洋强国),[161] and space great power (航天强国).[162]

As such, the PRC is positioning itself as a new shaping power to reform the governance of the strategic new frontiers. The CCP leadership has declared that China needs to increase its participation in the formulation of the rules that concern the governance of strategic new frontiers. For instance, former Politburo member Yang Jiechi has emphasised the importance of active involvement in international governance of the new frontiers by providing Chinese solutions,[163] while Foreign Minister Wang Yi has stressed the need to improve governance rules of the new frontiers so as to follow the global governance approach of extensive consultation, joint contribution, and shared benefits to prevent and resolve global security dilemmas.[164] Yang Jian maintains that the PRC's increasing national capabilities and moral values make its participation in global governance, including in the new frontiers, inevitable. According to Yang, the formation of a Community of Shared Future for Mankind serves as an obvious answer to address the deficiencies of the current governance system in the new frontiers. This is because the concept reflects the Chinese cultural idea of "harmonious coexistence," which considers overall and individual interests, immediate and long-term interests, and emphasises the sustainability of social development and the mutual support between humanity and nature. These core values, Yang contends, are essential to solving today's global problems, particularly in governing the new frontiers.[165]

However, to fully realise these ambitions, Chinese analysts agree that the PRC needs to further enhance its understanding, material capabilities, and international interactions in the strategic new frontiers. Knowledge, science, and technological development are seen as the foundation for exploration, utilisation, and governance in these domains.[166] Deng argues that states with knowledge power will hold the dominant position and discourse in establishing new rules and regulations for the new frontiers.[167] Military-civil fusion can play a role in developing such knowledge power. The *2020 Science of Military Strategy* states that, for instance, in the polar regions, military power and civilian power should be closely integrated, with military forces supporting polar scientific research by providing equipment, technology, medical assistance, and logistical support for national polar scientific expeditions and to further expand the ways and means of joint military and civilian activities.[168] The PRC's State Council, in its suggestions on promoting military-civil fusion in defence science, technology, and industry, states that it will speed up the coordinated development, sharing, and implementation of major military-civil fusion projects. These projects include focus areas, such as launch vehicles, remote sensing satellites, and sharing military and civilian satellite resources and data in the space domain; communication technology, network security, and the integration of land-orbit information networks in the cyber domain; and construction of deep-sea stations, ocean monitoring systems, underwater technologies (for detection and information transmission), development of icebreakers, polar rescue and resource exploration ships, polar semi-submersible transport ships and equipment in the maritime domain, which includes the polar regions.[169]

According to Chinese analysts, the PRC also needs to actively participate in international cooperation in the new domains and engage various stakeholders, including states and non-state actors, to prevent a single state or group of states from monopolising these domains.[170] It should engage scientists, non-governmental organisations, and important international organisations to jointly promote international governance in the new frontiers.[171] Chinese talent and experts are considered to play an important role in enhancing the PRC's international discourse power and competition over rulemaking in these domains.[172] According to Chinese analysts, this is because the development of global norms requires appealing ideas as well as persuasion and control over agenda-setting.[173]

It appears that the Chinese Dream of national rejuvenation, as envisioned by General Secretary Xi, is driving the PRC to establish a position of pre-eminence in East Asia and expand into regions beyond its immediate neighbourhood. Nadège Rolland convincingly argues that CCP elites aim to establish a partial, loose, and malleable hegemony. Partial because it prioritises spheres of influence over ruling the world, loose because it does not seek direct control over foreign territories, and malleable because there are no strict cultural, geographical, or ideological boundaries delineating Chinese hegemony – Asian as well as non-Asian countries can be included – as long as they defer to CCP primacy.[174] These ambitions are causing tensions between the PRC and the dominant power in the world, the US, as they now engage in a bipolar struggle for power. In this competition, the role of science and technology will be more prominent, and the civil-military divide will be blurred, making it more challenging for states to cooperate in areas that require significant science and technology input, such as the strategic new frontiers.

The PRC's Foreign Affairs Bureaucracy

General Secretary Xi has a large Party-state bureaucracy at his disposal to implement his vision of a rejuvenated China that commands international influence and is able to secure its interests around the world, including in the strategic new frontiers. The CCP, following Leninist principles of governing the state and society through tight centralisation of power, sits atop of this bureaucracy. The CCP penetrates all levels of the Chinese state, and almost all important officials, including State Council ministers, provincial leaders, military and diplomatic dignitaries, and CEOs of SOEs, are Party members.[175] The paramount leader and other senior officials at the top of the CCP, such as the Politburo and its Standing Committee, have historically played a crucial role in setting the PRC's overall foreign and security policy and managing international crises.[176] This role has become even more significant under Xi, who is considered the "prime mover" in China's foreign affairs,[177] and has strengthened top-down planning and coordination of the PRC's foreign and national security apparatus during his tenure.[178] Xi exercises authority through his leadership positions in several leading small groups and commissions, such as the Central Military Commission, the National Security Commission, and the Foreign Affairs Commission of the Central Committee. The CCP itself, through the International Department of the Central Committee, engages in PRC diplomacy by cultivating party-to-party relations and promoting the PRC's developmental initiatives and Chinese solutions worldwide.[179]

At a level below the Party bodies, several institutions, mostly at the ministerial level, are responsible for implementing policies set by the Party Centre and handling routine matters. In the realm of China's external relations, the Ministry of Foreign Affairs (MFA) is an important actor. It manages the PRC's day-to-day relations with foreign countries and serves as the primary point of contact for foreign governments and embassies in Beijing.[180] The MFA also has the authority to determine certain policies, particularly if the issue at hand is not deemed strategic.[181] It serves as a unique source of policy proposals, analysis, and information about the outside world.[182] However, due to the significant expansion of China's foreign interests in terms of geography and functionality over the past three decades (including trade, aid, science cooperation, military affairs, counter-terrorism, non-proliferation, education, etc.), the number of bureaucratic actors involved in Chinese foreign affairs has also increased. In addition to the MFA, other ministries and institutions represent the PRC internationally. For example, the Ministry of Ecology and Environment (MEE) participates in international climate change negotiations, the Ministry of Commerce (MOFCOM) is responsible for international trade negotiations and agreements, and the Ministry of State Security (MSS) engages in international intelligence collection and analysis.

Apart from these central and ministerial level organisations, many of China's international interactions, business deals, exchanges, and partnerships are conducted through Party-state-linked actors, including SOEs, research institutes, and provincial and municipal governments.[183] For example, Chinese SOEs, with their substantial capital resources and global presence, can conduct their own foreign visits or host official foreign delegations.[184] These SOEs and financial institutions are increasingly recognised as strategic actors capable of advancing the interests of the Chinese Party-state through their overseas projects, funding, and investments.[185] In matters related to strategic issues, such as energy security, Chinese decision-makers can and do consult specialists from SOEs.[186] By positioning themselves between projects in host countries and the PRC, these actors can manage a two-directional flow of information, knowledge, and expertise.

Chinese research institutions and think tanks are also considered sources of information and analysis that the Party-state leadership can utilise in its decision-making process. However, in the era of Xi Jinping, think-tank activities have expanded beyond traditional advisory functions and have become more prominent in China's public diplomacy.[187] General Secretary Xi has called for the development of new think tanks with Chinese characteristics to not only enhance the Party's governing abilities but also strengthen China's soft power.[188] As a result, Chinese think tanks and universities organise public events and international conferences and participate in high-level forums and summits to shape narratives on important global governance issues, particularly those involving China.[189] The PRC has been active in establishing global think-tank networks such as the Silk Road Think-Tank Network. According to some observers, these interactions enable PRC think tanks to gather information, engage with international media to shape foreign discourses about Chinese global activities in a manner favourable to the CCP, and cultivate present and future generations of elites with a positive view of the PRC.[190]

Finally, Chinese provinces and cities have also actively engaged in the PRC's foreign affairs by establishing numerous cooperative agreements and partnerships with foreign counterparts across the globe.[191] General Secretary Xi has highlighted the importance of subnational engagements in the PRC's relations with other major powers, which he considers the foundation for the development of bilateral relations.[192] Furthermore, provinces in China's border regions are actively positioning themselves as hubs for cross-border trade and engagement with neighbouring states, such as the Chinese southern provinces and Southeast Asian states.[193] In this capacity, they can also serve as platforms to advance the PRC's international initiatives, including the BRI.[194]

The Arctic in PRC's Foreign Affairs Bureaucracy

Since Xi announced in 2014 his desire to transform China into a polar great power and given the rising international profile of the Arctic, the CCP leadership has been increasingly recognising the region's strategic and policy importance.[195] However, beyond the Party Centre and its affiliated organisations,[196] several State Council ministries deal with Arctic issues.[197] One of the key organisations that coordinates the PRC's activities in the Arctic (and Antarctic) is the Ministry of Natural Resources (MNR). The MNR is responsible for drafting laws and regulations concerning the polar regions, formulating strategies for their development and the marine economy, as well as promoting military-civil fusion in the field of natural resources.[198] Importantly, under the MNR, the Chinese Arctic and Antarctic Administration (CAA) manages polar affairs, plans expeditions, and organises major polar research projects, among other responsibilities.[199] In addition to the MNR, other ministries involved in Arctic matters include the following:

- The MFA, which sends representatives to AC meetings and other conferences, prepares official statements and documents related to the Arctic and coordinates with Arctic and non-Arctic states on bilateral and multilateral issues regarding the Arctic.[200] Since 2016, the MFA has also appointed a Special Representative for Arctic Affairs.

- The Ministry of Transport (MOT) is focused on polar navigation, including the emerging Arctic shipping routes. Its Maritime Safety Administration has published guidelines in Chinese for sailing in the Arctic Ocean, including an updated version of the sailing manual for the Northeast Passage in 2022.[201] The Ministry also oversees the China Classification Society (CCS), which provides technical specifications and standards for ships, offshore installations, and related industrial products. The CSS helped formulate the PRC's Guidelines for Polar Water Operational Manual (极地水域操作手册编写指南) published in 2017.[202]
- The Ministry of Agriculture and Rural Affairs (MARA) is responsible, among other things, for the management of distant ocean fisheries and the negotiation and implementation of bilateral and multilateral fisheries agreements.[203] This includes the Arctic Ocean, as the Ministry plans to strengthen its scientific research on Arctic fishery resources during the 14th Five-Year Development Plan and actively participate in Arctic fishery affairs.[204]
- The Ministry of Science and Technology (MOST) is responsible for funding science and technology initiatives, organising basic research plans, major science and engineering projects, and promoting military-civil fusion in science and technology.[205] The Ministry is actively engaged in international polar science and technology cooperation at the bilateral levels (with partners such as Russia)[206] and multilateral levels (such as in polar-focused BRICS meetings).[207]

Other ministry-level organisations under the State Council are also involved in Arctic matters. The Ministry of Industry and Information Technology (MIIT) has a strong focus on telecommunication technologies, including cyberspace and network security.[208] It was previously engaged in preliminary negotiations for the development of a fibre-optic link through the Arctic Ocean.[209] The Ministry of Education (MOE) oversees universities conducting Arctic-related research, such as Tongji University, Ocean University of China, and Wuhan University.[210] The MEE is responsible for supervising polar environmental protection work,[211] while the National Energy Administration (NEA) participates in international energy cooperation, including with Arctic states like Russia.[212] Additionally, it is reasonable to assume that the Party-state's main intelligence-gathering agency, the MSS, is also interested in Arctic-related matters as there have been reports in recent years of apprehending foreign Arctic scientists spying for the PRC.[213] To coordinate the work of these organisations, an inter-ministerial coordination committee on Arctic issues was established in 2011, with the MFA leading the effort.[214] This mechanism later became a platform for the State Council ministries to draft the PRC's Arctic White Paper.[215] The handling of Arctic-related issues within the PRC's foreign affairs bureaucracy, beyond the ministerial level agencies, including the research institutes and the SOEs, will be discussed in subsequent chapters of the book.

Notes

1 "习近平：承前启后 继往开来 继续朝着中华民族伟大复兴目标奋勇前进" [Xi Jinping: Build on the Past Achievements and Forge Ahead Towards the Goal of National Rejuvenation], *Xinhua*, November 29, 2012, www.xinhuanet.com//politics/2012-11/29/c_113852724.htm.

2 For a short overview of national rejuvenation narratives of past and present Chinese leaders, see Friso M.S. Stevens, "China's Long March to National Rejuvenation: Toward a Neo-Imperial Order in East Asia?" *Asian Security* 17, no. 1 (2021): 46–63.

3 Elizabeth C. Economy, *The Third Revolution: Xi Jinping and the New Chinese State* (New York: Oxford University Press, 2018), 10.

4 Maria Adele Carrai, "Chinese Political Nostalgia and Xi Jinping's Dream of Great Rejuvenation," *International Journal of Asian Studies* 18 (2021): 12.

5 Bill Bishop, "China's Political Discourse November 2021: A New Resolution on History," *China Media Project*, December 30, 2021, https://sinocism.com/p/chinas-political-discourse-november.

6 CCP Leadership Group of the National Development and Reform Commission, "A Historic Leap Toward the Rejuvenation of the Chinese Nation," *Qiushi Journal*, September–October 2021, http://en.qstheory.cn/2021-11/15/c_680117.htm.

7 Lin Hui et al., "习近平的文化情怀" [Xi Jinping's Cultural Feelings], 人民日报 [*People's Daily*], May 12, 2022, http://paper.people.com.cn/rmrb/html/2022-05/12/nw.D110000renmrb_20220512_2-01.htm.

8 Camilla T.N. Sørensen, "The Significance of Xi Jinping's Chinese Dream for Chinese Foreign Policy: From Tao Guang Yang Hui to Fen Fa You Wei," *Journal of China and International Relations* 3, no. 1 (2015): 65.

9 Ashley J. Tellis, "Pursuing Global Reach: China's Not So Long March Toward Preeminence," in *China's Expanding Strategic Ambitions*, edited by Ashley J. Tellis, Alison Szalwinski and Michael Wills (Seattle and Washington, DC: The National Bureau of Asian Research, 2019), 29.

10 Meng Xiangqing, "Chinese Dream Includes Strong PLA," *China Daily*, October 8, 2013, www.china.org.cn/opinion/2013-10/08/content_30223195.htm.

11 Xi Jinping, "Secure a Decisive Victory in Building a Moderately Prosperous Society in All Respects and Strive for the Great Success of Socialism with Chinese Characteristics for a New Era," *Xinhua*, October 18, 2017, www.xinhuanet.com/english/download/Xi_Jinping's_report_at_19th_CPC_National_Congress.pdf.

12 Han Qingxiang and Chen Shuguan, "中华民族伟大复兴的世界意义" [The Global Significance of the Great Rejuvenation of the Chinese Nation], 人民网 [*People's Daily Online*], May 5, 2016, http://theory.people.com.cn/n1/2016/0505/c40531-28325921.html.

13 Wang Yi, "Striding Forward Holding High the Banner of Building a Community with a Shared Future for Mankind," *Ministry of Foreign Affairs of the PRC*, January 1, 2022, www.fmprc.gov.cn/mfa_eng/wjdt_665385/zyjh_665391/202201/t20220101_10478338.html.

14 Rush Doshi, *The Long Game: China's Grand Strategy to Displace American Order* (New York: Oxford University Press, 2021), 286.

15 Li Xiang, "High-Quality Development Tops Agenda," *China Daily*, November 25, 2021, www.chinadaily.com.cn/a/202111/25/WS619ec9d7a310cdd39bc775be.html.

16 "Made in China 2025 Plan Issued," *The State Council of the PRC*, May 19, 2015, http://english.www.gov.cn/policies/latest_releases/2015/05/19/content_281475110703534.htm. For a detailed description of the plan in Chinese see "国务院关于印发《中国制造2025》的通知" [State Council Notice About Issuing Made in China 2025], 中华人民共和国中央人民政府 [*The State Council of the PRC*], May 19, 2015, www.gov.cn/zhengce/content/2015-05/19/content_9784.htm.

17 Elizabeth C. Economy, *The World According to China* (Cambridge: Polity Press, 2022), 145. Although the Chinese government refrained from mentioning the plan in

its recent statements due to international criticism, Economy notes that the program lives on under the rubric "industrial upgrading." See Ibid., 147.

18 Alicia García Herrero, "What Is Behind China's Dual Circulation Strategy," *China Leadership Monitor* 69 (2021), www.prcleader.org/herrero.

19 Jude Blanchette and Andrew Polk, "Dual Circulation and China's New Hedged Integration Strategy," *CSIS*, August 24, 2020, www.csis.org/analysis/dual-circulation-and-chinas-new-hedged-integration-strategy.

20 Nis Grünberg and Vincent Brussee, "China's 14th Five-Year Plan – Strengthening the Domestic Base to Become a Superpower," *MERICS*, April 9, 2021, https://merics.org/en/comment/chinas-14th-five-year-plan-strengthening-domestic-base-become-super-power.

21 Wang Zheng, "制造业正从中国制造向中国创造迈进" [The Manufacturing Industry Is Moving from Made in China to Developed in China], 人民日报 [*People's Daily*], March 21, 2022, http://paper.people.com.cn/rmrb/html/2022-03/21/nw.D110000renmrb_20220321_5-01.htm.

22 "工业和信息化部关于印发"十四五"大数据产业发展规划的通知" [A Notice of the Ministry of Industry and Information Technology on the Publishing of the 14th Five-Year Plan for Big Data Industry Development], 工业和信息化部 [*Ministry of Industry and Information Technology of the PRC*], November 15, 2021, www.miit.gov.cn/zwgk/zcwj/wjfb/tz/art/2021/art_c4a16fae377f47519036b26b474123cb.html.

23 Liu Yuying, "国家标准委：正制定《中国标准2035》" [Standardization Administration of China: Formulating the China Standards 2035], 中国新闻网 [*China News*], January 10, 2018, https://china.huanqiu.com/article/9CaKrnK6iOH.

24 Arjun Gargeyas, "China's Standards 2035 Project Could Result in a Technological Cold War," *The Diplomat*, September 18, 2021, https://thediplomat.com/2021/09/chinas-standards-2035-project-could-result-in-a-technological-cold-war/.

25 Ibid.

26 Xi Jinping, "努力成为世界主要科学中心和创新高地" [Strive to Become the World's Leading Scientific Centre and Innovation Hub], 求是 [*Qiushi*], March 15, 2021, www.xinhuanet.com/2021-03/15/c_1127212833.htm.

27 Zhao Yongxin, "深入实施创新驱动发展战略：我国成功进入创新型国家行列" [Deeply Implement the Innovation-Driven Development Strategy: China Successfully Entered the Ranks of Innovative Countries], 人民日报 [*People's Daily*], June 7, 2022, http://paper.people.com.cn/rmrb/html/2022-06/07/nw.D110000renmrb_20220607_2-02.htm.

28 For example, see Ryan Martinson, "Gliders with Ears: A New Tool in China's Quest for Undersea Security," *Centre for International Maritime Security*, March 21, 2022, https://cimsec.org/gliders-with-ears-a-new-tool-in-chinas-quest-for-undersea-security/.

29 Han Jie et al., "中国经济发展前景一定会更加光明" [The Prospects for China's Economic Development Will Be Even Brighter], 人民日报 [*People's Daily*], May 25, 2022, https://paper.people.com.cn/rmrb/html/2022-05/25/nw.D110000renmrb_20220525_1-01.htm.

30 Ren Ping, "全面塑造发展新优势 – 论坚持创新在我国现代化建设全局中的核心地位" [Comprehensively Develop New Advantages for Development – On Persisting in the Core Position of Innovation in China's Overall Modernization Construction], 人民日报 [*People's Daily*], October 28, 2021, http://paper.people.com.cn/rmrb/html/2021-10/28/nw.D110000renmrb_20211028_1-05.htm.

31 Ibid.

32 Doshi, *The Long Game: China's Grand Strategy to Displace American Order*, 287.

33 Jiang Jibao, "百年未有之大变局 中美博弈谁主沉浮" [Changes Unseen in a Century – Who Will Win and Who Will Lose in the Sino-American Competition], 中国青年报 [*China Youth Daily*], August 13, 2019, https://zqb.cyol.com/html/2019-08/13/nw.D110000zgqnb_20190813_4-03.htm.

34 Wu Jingjing and Yu Xiaojie, "《国家创新驱动发展战略纲要》印发 提出2050年建成世界科技创新强国" [Outline of the National Strategy of Innovation-Driven Development Issued, Proposes That China Will Become a World S&T Innovation New

Great Power by 2050], *Xinhua*, May 20, 2016, www.gov.cn/xinwen/2016-05/20/content_5074905.htm.

35 For the purposes of this research, referring to Chinese R&D will point to basic and applied research, adopted from Alex Stone, *China's Model of Science: Rationale, Players, Issues* (Montgomery, AL: China Aerospace Studies Institute, 2022), 13.

36 Cheng Yu, "Country's Spending on R&D Reaches Record High in 2021," *China Daily*, January 27, 2022, https://global.chinadaily.com.cn/a/202201/27/WS61f1d7a2a310cd-d39bc83765.html.

37 Shawn Kim, "China Standards 2035: How China Plans to Win the Future with Its Own International Tech Standards," *South China Morning Post*, May 21, 2021, www.scmp.com/comment/opinion/article/3134216/china-standards-2035-how-china-plans-win-future-its-own.

38 Ren, "全面塑造发展新优势 – 论坚持创新在我国现代化建设全局中的核心地位" [Comprehensively Develop New Advantages for Development – On Persisting in the Core Position of Innovation in China's Overall Modernization Construction].

39 Research Group of the Institute of International and Strategic Studies at Peking University, "技术领域的中美战略竞争：分析与展望" [Sino-American Strategic Competition in Technology: Analysis and Prospect], 北京大学国际战略研究院 [*Institute of International and Strategic Studies at Peking University*], January 30, 2022, https://web.archive.org/web/20220629133257/http://cn3.uscnpm.org/model_item.html?action=view&table=article&id=27016.

40 For example, see Jeffrey Stoff, "China's Talent Programs," in *China's Quest for Foreign Technology: Beyond Espionage*, eds. William C. Hannas and Didi Kirsten Tatlow (London and New York: Routledge, 2021), 38–54.

41 "Xi Focus: Xi Calls for Accelerating Building of World Centre for Talent, Innovation," *Xinhua*, September 28, 2022, www.news.cn/english/2021-09/28/c_1310215793.htm.

42 Sylvia Schwaag Serger et al., "What Do China's Scientific Ambitions Mean for Science – and the World?" *Issues in Science and Technology*, April 5, 2021, https://issues.org/what-do-chinas-scientific-ambitions-mean-for-science-and-the-world/.

43 See, for example, Xi Jinping, "在中国科学院第二十次院士大会、中国工程院第十五次院士大会、中国科协第十次全国代表大会上的讲话" [Speech at the 20th Conference of Academicians of the Chinese Academy of Sciences, the 15th Conference of Academicians of the Chinese Academy of Engineering, and the 10th National Congress of the China Association for Science and Technology], *Xinhua*, May 28, 2021, www.xinhuanet.com/politics/leaders/2021-05/28/c_1127505377.htm, Cheng Yingqi, "Scientists Called to Serve," *China Daily*, July 18, 2013, www.chinadaily.com.cn/china/2013-07/18/content_16791062.htm.

44 Serger et al., "What Do China's Scientific Ambitions Mean for Science – and the World?"

45 Stone, *China's Model of Science: Rationale, Players, Issues*, 21.

46 Ibid., 20–22. See also Wu Yuehui et al., "坚持和加强党对科技事业的领导" [Upholding and Strengthening the Party's Leadership Over Science and Technology], 人民网 [*People's Daily Online*], June 2, 2018, http://politics.people.com.cn/n1/2018/0602/c1001-30029995.html.

47 Serger et al., "What Do China's Scientific Ambitions Mean for Science – and the World?"

48 Stone, *China's Model of Science: Rationale, Players, Issues*, 7. See also "中共中央政治局召开会议 审议《国家安全战略（2021－2025年）》《军队功勋荣誉表彰条例》和《国家科技咨询委员会2021年咨询报告》 中共中央总书记习近平主持会议" [The Politburo of the CCP Central Committee Held a Meeting to Review the "National Security Strategy (2021–2025)," "Regulations on Commendation of Military Merit and Honours" and "2021 Consultation Report of the National Science and Technology Advisory Committee" General Secretary of the CCP Xi Jinping Presided Over the Meeting], *Xinhua*, November 18, 2021, http://cpc.people.com.cn/n1/2021/1118/c64094-32286177.html.

49 Zhao Yongxin and Gu Yekai, "推进科技政策扎实落地 – 访科技部党组书记、部长王志刚" [Promoting the Implementation of Science and Technology Policies – Interview with

Wang Zhigang, Party Secretary at the Ministry of Science and Technology and the Ministry of Science and Technology], 人民日报 [*People's Daily*], December 23, 2021, http://paper.people.com.cn/rmrb/html/2021-12/23/nw.D110000renmrb_20211223_3-02.htm.

50 "Xi: National Security a Matter of Prime Importance," *Xinhua*, April 16, 2014, www.china.org.cn/china/2014-04/16/content_32104039.htm.

51 Daniel Tobin, "World Class: The Logic of China's Strategy and Global Military Ambitions," in *Securing the China Dream: The PLA's Role in a Time of Reform and Change*, eds. Roy Kamphausen, David Lai and Tiffany Ma (Seattle: The National Bureau of Asian Research, 2020), 30.

52 "Full Text: China's National Defence in the New Era," *The State Council Information Office of the PRC*, July 24, 2019, http://english.scio.gov.cn/2019-07/24/content_75026800.htm.

53 Tellis, "Pursuing Global Reach: China's Not So Long March toward Pre-eminence," 35.

54 Jean-Pierre Cabestan, "China's Military Base in Djibouti: A Microcosm of China's Growing Competition with the United States and New Bipolarity," *Journal of Contemporary China* 29, no. 125 (2020): 731–747. There have also been reports that the PLA was trying to establish overseas bases in West Africa, the United Arab Emirates and Cambodia, see The Editorial Board, "The Chinese Navy's Great Leap Forward," *The Wall Street Journal*, June 7, 2022, www.wsj.com/articles/chinas-great-naval-leap-forward-cambodia-military-base-navy-beijing-xi-jinping-11654637895.

55 Gordon Lubold, "Advanced Manoeuvre in China Hypersonic Missile Test Shows New Military Capability," *The Wall Street Journal*, November 21, 2021, www.wsj.com/articles/advanced-maneuver-in-china-hypersonic-missile-test-shows-new-military-capability-11637545843.

56 Charles A. Richard, "Statement of Charles A. Richard Commander United States Strategic Command Before the House Appropriations Subcommittee on Defence 5 April 2022," *The House Committee on Appropriations*, April 5, 2022, https://docs.house.gov/meetings/AP/AP02/20220405/114575/HHRG-117-AP02-Wstate-RichardC-20220405.pdf.

57 Tellis, "Pursuing Global Reach: China's Not So Long March toward Pre-eminence," 39–40.

58 Diego Lopes Da Silva et al., "Trends in World Military Expenditure, 2021," *SIPRI Fact Sheet* (April 2022): 4.

59 "Full Text: China's National Defence in the New Era," *The State Council Information Office of the PRC*.

60 Doshi, *The Long Game: China's Grand Strategy to Displace American Order*, 293.

61 David Vergun, "Freedom of Navigation in South China Sea Critical to Prosperity, Says Indo-Pacific Commander," *US Department of Defence*, November 23, 2019, www.defense.gov/News/News-Stories/Article/Article/2025105/freedom-of-navigation-in-south-china-sea-critical-to-prosperity-says-indo-pacif/.

62 See, Ronald O'Rourke, *China Naval Modernization: Implications for US Navy Capabilities – Background and Issues for Congress* (Washington, DC: Congressional Research Service, 2021).

63 Ibid., 3.

64 Office of the Secretary of Defence, *Annual Report to Congress: Military and Security Developments Involving the People's Republic of China 2020* (Washington, DC: Department of Defence, 2020).

65 H.I. Sutton, "Chinese Navy Growth: Massive Expansion of Important Shipyard," *Naval News*, March 15, 2022, www.navalnews.com/naval-news/2022/03/chinese-navy-growth-massive-expansion-of-important-shipyard/.

66 See Ryan D. Martinson and Peter A. Dutton, "China Maritime Report No. 3: China's Distant-Ocean Survey Activities: Implications for US National Security," *CMSI China Maritime Reports* 3 (2018): 1–32.

67 Isaac B. Kardon and Wendy Leutert, "Pier Competitor: China's Power Position in Global Ports," *International Security* 46, no. 4 (2022): 9–47.

68 Zhu Zihua, "从海洋大国到海洋强国" [From a Maritime Power to a Maritime Great Power], 解放军报 [*PLA Daily*], April 7, 2015, www.81.cn/jwgd/2015-04/07/content_6431231.htm.

69 Thomas Bickford, "China and Maritime Power: Meanings, Motivations, and Strategy," in *Becoming a Great Maritime Power: A Chinese Dream*, ed. Michael McDevitt (Arlington: CAN Analysis and Solutions, 2016), 7–8.

70 Li Jian, "新时代呼唤新的国家海洋观" [A New Era Calls for a New National Ocean Outlook], 解放军报 [*PLA Daily*], April 25, 2017, http://theory.people.com.cn/n1/2017/0425/c40531-29233362.html.

71 Toshi Yoshihara and James R. Holmes, *Red Star Over the Pacific: China's Rise and the Challenge to US Maritime Strategy*, second edition (Annapolis: Naval Institute Press, 2018), 5.

72 "深入把握总体国家安全观" [Fully Grasp the Comprehensive National Security Outlook], 人民网 [*People's Daily Online*], November 22, 2021, http://theory.people.com.cn/n1/2021/1122/c148980-32288411.html.

73 "总体国家安全观的16种安全" [16 Kinds of Security of the Comprehensive National Security Outlook], 国安宣工作室 [*National Security Propaganda Office*], April 14, 2021, www.stdaily.com/cehua/20210414/2021-04/14/content_1114342.shtml.

74 Centre for the Study of Comprehensive National Security, "深刻认识中国特色国家安全道路" [Understand Profoundly the National Security Path with Chinese Characteristics], 人民日报 [*People's Daily*], April 15, 2022, http://paper.people.com.cn/rmrb/html/2022-04/15/nw.D110000renmrb_20220415_1-09.htm.

75 Timothy Heath, "The Holistic Security Concept: The Securitization of Policy and Increasing Risk of Militarized Crisis," *China Brief* 15, no. 12 (2015), https://jamestown.org/program/the-holistic-security-concept-the-securitization-of-policy-and-increasing-risk-of-militarized-crisis/#.VY1dXOs4SS0.

76 "习近平：坚持总体国家安全观 走中国特色国家安全道路" [Xi Jinping: Adhere to the Holistic National Security Concept and Follow the Path of National Security with Chinese Characteristics], *Xinhua*, April 15, 2014, www.xinhuanet.com/politics/2014-04/15/c_1110253910.htm.

77 Yuan Peng, "以高水平安全保障高质量发展" [Ensure High-Quality Development with High-Level Security], 人民日报 [*People's Daily*], January 5, 2022, http://theory.people.com.cn/n1/2022/0105/c40531-32324058.html.

78 Zhong Yin, "树牢总体国家安全观" [Build Up the Overall National Security Concept], 人民日报 [*People's Daily*], April 15, 2022, http://paper.people.com.cn/rmrb/html/2022-04/15/nw.D110000renmrb_20220415_2-02.htm.

79 Heath, "The Holistic Security Concept: The Securitization of Policy and Increasing Risk of Militarized Crisis."

80 For a comprehensive and in-depth analysis of the CCP's MCF see Alex Stone and Peter Wood, *China's Military-Civil Fusion Strategy: A View from Chinese Strategists* (Montgomery, AL: China Aerospace Studies Institute, 2020).

81 Jin Zhuanglong, "开创新时代军民融合深度发展新局面" [Open Up New Horizons for Deepening of the MCF in the New Era], 求是 [*Qiushi*], July 16, 2018, www.xinhuanet.com/politics/2018-07/16/c_1123133733.htm.

82 "China's Military Strategy (2015)," *The State Council of the PRC*, May 27, 2015, http://english.www.gov.cn/archive/white_paper/2015/05/27/content_281475115610833.htm.

83 "中华人民共和国国民经济和社会发展第十四个五年规划和2035年远景目标纲要" [Outline of the 14th Five-Year Plan (2021–2025) for National Economic and Social Development and Vision 2035 of the People's Republic of China], *Xinhua*, March 13, 2021, www.gov.cn/xinwen/2021-03/13/content_5592681.htm.

84 Audrey Fritz, "The Foundation of Innovation Under Military-Civil Fusion: The Role of Universities," *Sinopsis*, October 8, 2021, https://sinopsis.cz/wp-content/uploads/2021/10/mcf0.pdf.

85 Wang Lu, "推动军民融合深度发展" [Promote the Development of Military-Civil Fusion], 人民网 [*People's Daily Online*], November 21, 2017, http://theory.people. com.cn/n1/2017/1121/c40531-29657926.html.

86 Stone and Wood, *China's Military-Civil Fusion Strategy: A View from Chinese Strategists*, 36.

87 Jiang Luming, "为何把军民融合上升为国家战略" [Why the MCF Was Elevated to a National Strategy], 人民日报 [*People's Daily*], September 25, 2017, www.aisixiang. com/data/106149.html.

88 Matthew P. Funaiole, Brian Hart and Joseph S. Bermudez Jr., "In the Shadow of Warships: How foreign Companies Help Modernize China's Navy," *CSIS*, n.d., https:// features.csis.org/china-shadow-warships/.

89 Brendan S. Mulvaney, "Preface," in *China's Military-Civil Fusion Strategy: A View from Chinese Strategists*, eds. Alex Stone and Peter Wood (Montgomery, AL: China Aerospace Studies Institute, 2020).

90 Greg Levesque and Mark Stokes, "Blurred Lines: Military-Civil Fusion and the Going Out of China's Defence Industry," *Pointe Bello Report*, December 2016, https://web.archive.org/ web/20200505042315/https://static1.squarespace.com/static/569925bfe0327c837e2e9a94/ t/593dad0320099e64e1ca92a5/1497214574912/062017_Pointe+Bello_ Military+Civil+Fusion+Report.pdf.

91 In particular, there are seven defence research universities directly involved in the MCF, often referred to as the "Seven sons of National Defence" (国防七子). These include North-western Polytechnical University, Harbin Engineering University, Harbin Institute of Technology, Beihang University, Beijing Institute of Technology, Nanjing University of Science and Technology, and Nanjing University of Aeronautics and Astronautics, see Glenn Tiffert (ed.), *Global Engagement: Rethinking Risk in the Research Enterprise* (Stanford: Hoover Institution Press, 2020).

92 Zeng Jinghan, Xiao Yuefan and Shaun Breslin, "Securing China's Core Interests: The State of the Debate in China," *International Affairs* 91, no. 2 (2015): 245–266.

93 Ling Wei, "Striving for Achievement in a New Era: China Debates Its Global Role," *The Pacific Review* 33 (2020): 418–419.

94 Sørensen, "The Significance of Xi Jinping's Chinese Dream for Chinese Foreign Policy: From Tao Guang Yang Hui to Fen Fa You Wei," 66.

95 Yan Xuetong, "From Keeping a Low Profile to Striving for Achievement," *The Chinese Journal of International Politics* 7, no. 2 (2014): 166.

96 Li Juan, "塑造力彰显中国特色大国外交的担当" [Shaping Power Demonstrates China's Major Power Diplomacy Responsibility], 文汇报 [*Wen Wei Po*], December 12, 2017, www.china.com.cn/opinion/theory/2017-12/12/content_41982208.htm.

97 Nadège Rolland, *China's Vision for a New World Order* (Seattle: The National Bureau of Asian Research, 2020), 13–14.

98 For a comprehensive overview of the concept of "great changes unseen in a century" see Doshi, *The Long Game: China's Grand Strategy to Displace American Order*, 265–271. For a Chinese leadership source, see "习近平接见2017年度驻外使节工作会议与会使节并发表重要讲话" [Xi Jinping Met the 2017 Ambassadorial Conference and Delivered an Important Speech], *Xinhua*, December 28, 2017, www.xinhua-net.com/politics/leaders/2017-12/28/c_1122181743.htm.

99 Zhou Li, "当前世界主要矛盾与国际格局演变" [Major Contradictions in the World and the Evolution of the International Structure], 经济导刊 [*Economic Herald*], February 23, 2022, www.jingjidaokan.com/icms/null/null/ns:LHQ6LGY6LGM6MmM5ZTg1YT Q3ZTMzOGNiMDAxN2YyNDU2MTE3ZDAwNDYscDosYTosbTo=/show.vsml.

100 "Full Text: Chinese President Xi Jinping's Keynote Speech at the Opening Ceremony of BFA Annual Conference 2022," *Xinhua*, April 21, 2022, https://english.news. cn/20220421/f5f48ba605ed427dab911188af175ebf/c.html.

101 Chen Zhimin, "真正的多边主义的理论内涵" [Theoretical Contents of True Multilateralism], 人民日报 [*People's Daily*], January 17, 2022, http://paper.people.com.cn/ rmrb/html/2022-01/17/nw.D110000renmrb_20220117_1-09.htm.

102 "Xi Stresses Urgency to Reform Global Governance," *Xinhua*, October 13, 2015, www.china.org.cn/china/2015-10/13/content_36805468.htm.

103 Chen Xiangjian and Zhang Wenbiao, "新时代中国特色大国外交的根本遵循" [The Fundamental Principles of Major Country Diplomacy with Chinese Characteristics], 红旗文稿 [*Hong Qi Wen Gao*], January 26, 2022, www.qstheory.cn/dukan/hqwg/2022-01/26/c_1128302520.htm.

104 Zhao Kejin, "新时代中国国际战略探索" [Exploring China's Global Strategy in the New Era], 国际问题研究 [*China International Studies*] 5 (2021): 22.

105 "China and the World in the New Era," *The State Council Information Office of the PRC*, September 28, 2019, http://english.scio.gov.cn/2019-09/28/content_75252746_2.htm.

106 "Vision and Actions on Jointly Building Silk Road Economic Belt and 21st-Century Maritime Silk Road," *Xinhua*, March 30, 2015, https://eng.yidaiyilu.gov.cn/qwyw/qwfb/1084.htm.

107 "马拉维加入"一带一路", 成第149个签署"一带一路"文件国家" [Malawi Joined the Belt and Road Initiative Becoming the 149th Country to Sign the Document], 北京日报 [*Beijing Daily*], March 31, 2022, http://fec.mofcom.gov.cn/article/fwydyl/zgzx/202203/20220303301095.shtml.

108 Rolland, *China's Vision for a New World Order*, 41.

109 Hu Jianguo, "一带一路是战略构想不是工程项目" [The Belt and Road Is a Strategic Concept, Not an Engineering Project], 中国经济网 [*China Economic Net*], March 19, 2015, http://finance.people.com.cn/money/n/2015/0319/c42877-26719589.html.

110 Wang Yi, "落实全球安全倡议, 守护世界和平安宁" [Implement the Global Security Initiative, Safeguard World Peace and Stability], 人民日报 [*People's Daily*], April 24, 2022, http://paper.people.com.cn/rmrb/html/2022-04/24/nw.D110000renmrb_20220424_1-06.htm.

111 Han Liang et al., "胸怀天下谋大同--习近平主席倡导的全球治理观深刻启迪世界" [Cherish the World and Seek Great Harmony – Xi Jinping's Vision of Global Governance Inspired the World], 人民日报 [*People's Daily*], June 21, 2022, http://paper.people.com.cn/rmrb/html/2022-06/21/nw.D110000renmrb_20220621_3-01.htm#.

112 Zhang Guihong, "国际组织与一带一路的多边议程化" [International Organizations and the BRI's Multilateral Agenda], 中国社会科学报 [*Chinese Social Sciences Today*], March 19, 2022, www.aisixiang.com/data/132116.html.

113 Doshi, *The Long Game: China's Grand Strategy to Displace American Order*, 282–283.

114 "金砖国家领导人第十四次会晤举行" [The 14th BRICS Summit Was Held], 人民日报 [*People's Daily*], June 24, 2022, https://paper.people.com.cn/rmrb/html/2022-06/24/nw.D110000renmrb_20220624_1-01.htm.

115 "Joint Statement of the Russian Federation and the People's Republic of China on the International Relations Entering a New Era and the Global Sustainable Development," *President of Russia*, February 4, 2022, http://en.kremlin.ru/supplement/5770.

116 Vasily Kashin, "Russia and China Take Military Partnership to New Level," *The Moscow Times*, October 23, 2019, www.themoscowtimes.com/2019/10/23/russia-and-china-take-military-partnership-to-new-level-a67852.

117 Wei Lingling and Hua Sha, "China's Xi Reaffirms Support for Moscow in Call with Putin," *The Wall Street Journal*, June 15, 2022, www.wsj.com/articles/chinas-xi-fails-to-endorse-putin-over-ukraine-in-call-with-russian-leader-11655299293.

118 Rolland regards the various Chinese initiatives (the BRI in particular) and the Community of Shared Future as two sides of the same coin: the Community being an intellectual construct – a theory, while the BRI is the practice, or the part, that knits the Community together. See Nadège Rolland, "Beijing's Vision for a Reshaped International Order," *China Brief* 18, no. 3 (2018), https://jamestown.org/program/beijings-vision-reshaped-international-order/.

119 Party Committee of the Ministry of Foreign Affairs, "以习近平外交思想为引领开创新时代外交工作新局面" [Open Up New Prospects for China's Diplomacy in

the New Era Under the Guidance of Xi Jinping Thought on Diplomacy], 人民日报 [*People's Daily*], December 7, 2021, http://paper.people.com.cn/rmrb/html/2021-12/07/nw.D110000renmrb_20211207_1-09.htm.

120 Zhao Xiaochun, "In Pursuit of a Community of Shared Future: China's Global Activism in Perspective," *China Quarterly of International Strategic Studies* 4, no. 1 (2018): 27.

121 Xi Jinping, "Work Together to Build a Community of Shared Future for Mankind," *Xinhua*, January 19, 2017, www.xinhuanet.com/english/2017-01/19/c_135994707.htm.

122 "China and the World in the New Era," *The State Council Information Office of the PRC*.

123 Yang Jiechi, "推动构建人类命运共同体" [Promoting the Building of a Community of Shared Future for Mankind], 人民日报 [*People's Daily*], November 26, 2021, www.aisixiang.com/data/129920.html.

124 Rolland, *China's Vision for a New World Order*, 38–39.

125 Ibid., 7.

126 "习近平在中共中央政治局第三十次集体学习时强调　加强和改进国际传播工作 展示真实立体全面的中国" [During the 30th Collective Study Session of the Political Bureau of the Central Committee of the CCP, Xi Jinping Emphasized Strengthening and Improving International Communication Work to Demonstrate a True, Three-Dimensional and Comprehensive China], *Xinhua*, June 1, 2021, www.xinhuanet.com/politics/2021-06/01/c_1127517461.htm.

127 Hu Zhengrong, "新时代中国国际话语权建构的现状与进路" [The Current Situation and Approach of the Construction of China's International Discourse Power in the New Era], 人民论坛 [*People's Tribune*], January 29, 2022, www.rmlt.com.cn/2022/0129/639052.shtml.

128 Rolland, *China's Vision for a New World Order*, 12.

129 He Yiting, "中华民族伟大复兴与中国话语的崛起" [The Great Rejuvenation of the Chinese Nation and the Rise of Chinese Discourse], 学习时报 [*Study Times*], September 27, 2019, www.qstheory.cn/llwx/2019-09/27/c_1125046449.htm.

130 Rolland, *China's Vision for a New World Order*, 28.

131 "Xi Focus: Xi Calls for Blazing New Path to Develop China's World-Class Universities," *Xinhua*, April 25, 2022, https://english.news.cn/20220425/0a0c73e26bf74cfea196bcebfde8c7ae/c.html.

132 Xie Fuzhan, "建构中国自主的知识体系" [Constructing China's Independent Knowledge System], 人民日报 [*People's Daily*], May 17, 2022, http://paper.people.com.cn/rmrb/html/2022-05/17/nw.D110000renmrb_20220517_1-09.htm.

133 Daniel C. Lynch, "Is China's Rise Now Stalling?" *The Pacific Review* 32, no. 3 (2019): 451–456.

134 See, for example, Roland Rajah and Alyssa Leng, "Revising Down the Rise of China," *Lowy Institute*, March 14, 2022, www.lowyinstitute.org/publications/revising-down-rise-china.

135 Logan Wright, "Rethinking China's Economic Future," *Rhodium Group*, May 31, 2022, https://rhg.com/research/rethinking-chinas-economic-future/#:~:text=China's%20economy%20is%20clearly%20contracting,modest%20decline%20in%20industrial%20output.

136 Laura Silver, Christine Huang and Laura Clancy, "Negative Views of China Tied to Critical Views of Its Policies on Human Rights," *Pew Research Centre*, June 29, 2022, www.pewresearch.org/global/2022/06/29/negative-views-of-china-tied-to-critical-views-of-its-policies-on-human-rights/.

137 Rajeswari Pillai Rajagopalan, "The Growing Tech Focus of the Quad," *The Diplomat*, July 9, 2022, https://thediplomat.com/2022/07/the-growing-tech-focus-of-the-quad/.

138 David Ljunggren and Steve Scherer, "Canada to ban Huawei/ZTE 5G Equipment, Joining Five Eyes Allies," *Reuters*, May 20, 2022, www.reuters.com/business/media-telecom/canada-announce-ban-use-huawei-zte-5g-equipment-source-2022-05-19/.

139 Jude Blanchette, "Xi's Confidence Game," *Foreign Affairs*, November 23, 2021, www.foreignaffairs.com/articles/asia/2021-11-23/xis-confidence-game.

140 Zhang Xudong et al., "沿着必由之路夺取新的更大胜利" [Follow the Inevitable Path and Seize New and Greater Victories], 人民日报 [*People's Daily*], March 16, 2022, http://paper.people.com.cn/rmrb/html/2022-03/16/nw.D110000renmrb_20220316_3-01.htm.

141 Ma Jiantang and Zhao Changwen, "党领导人民创造了人类文明新形态" [The Party Led the People to Create a New Form of Human Civilization], 人民日报 [*People's Daily*], February 9, 2022, http://paper.people.com.cn/rmrb/html/2022-02/09/nw.D110000renmrb_20220209_1-07.htm.

142 Zhang Xudong et al., "用好有利条件 走好必由之路" [Make Good Use of Favourable Conditions and Take the Inevitable Route], 人民日报 [*People's Daily*], March 21, 2022, http://paper.people.com.cn/rmrb/html/2022-03/21/nw.D110000renmrb_20220321_1-01.htm.

143 Le Yucheng, "变乱交织的世界与勇毅前行的中国外交" [Chinese Diplomacy Forges Ahead Bravely in a Chaotic and Intertwined World], 中华人民共和国外交部 [Ministry of Foreign Affairs of the PRC], January 18, 2022, www.mfa.gov.cn/wjbxw_new/202201/t20220118_10629765.shtml.

144 "Xi Focus: Xi Stresses Good Start for Fully Building Modern Socialist China," *Xinhua*, January 12, 2021, www.xinhuanet.com/english/2021-01/12/c_139659544.htm.

145 Wen Yan, "深入学习习近平外交思想，努力开创中国特色大国外交新局面" [Deeply Study Xi Jinping's Thought on Diplomacy, Strive to Open up New Prospects for Major-Country Diplomacy with Chinese Characteristics], 人民网 [*People's Daily Online*], January 6, 2020, http://theory.people.com.cn/n1/2020/0106/c40531-31535409.html.

146 Huang Xiaoxi and Cui Qingxin, "国家安全法草案拟增加太空等新型领域的安全维护任务" [The Draft National Security Law Plans to Increase Security Safeguarding Tasks in New Domains Such as Outer Space], *Xinhua*, June 24, 2015, www.gov.cn/xinwen/2015-06/24/content_2883509.htm.

147 "中华人民共和国国家安全法" [The National Security Law of the People's Republic of China], 中华人民共和国国防部 [*Ministry of National Defence of the PRC*], April 14, 2016, www.81.cn/2016gjaqr/2016-04/14/content_7007225_2.htm.

148 Patrik Andersson, "The Arctic as a 'Strategic' and 'Important' Chinese Foreign Policy Interest: Exploring the Role of Labels and Hierarchies in China's Arctic Discourses," *Journal of Current Chinese Affairs* (2021): 17.

149 Zhang Zhijun and Liu Huirong, "当前国际法跨学科人才培养的新任务新课题 – 基于深海、极地、外空、网络等战略新疆域的思考" [New Tasks and Topics for Current International Law Interdisciplinary Talent Training – Based on Deep Sea, Polar Regions, Outer Space, Cyber Space and Other Strategic New Frontiers], 学术前沿 [*Frontiers*], no. 3 (2021), https://aoc.ouc.edu.cn/2022/0303/c9824a363730/pagem.htm.

150 Xiao Tianliang et al., *The Science of Military Strategy 2020*, trans. China Aerospace Studies Institute (Montgomery, AL: China Aerospace Studies Institute, 2022), 142.

151 Freeman's analysis focused particularly on international regimes that govern these domains, Carla P. Freeman, "An Uncommon Approach to the Global Commons: Interpreting China's Divergent Positions on Maritime and Outer Space Governance," *The China Quarterly* 241 (2020): 4.

152 Deng Beixi, "全球公域视角下的极地安全问题与中国的应对" [Analysis of Polar Security Issues from the Perspective of Global Commons and China's Countermeasures], 江南社会学院学报 [*Journal of Jiangnan Social University*] 20, no. 3 (2018): 35.

153 Andersson, "The Arctic as a 'Strategic' and 'Important' Chinese Foreign Policy Interest: Exploring the Role of Labels and Hierarchies in China's Arctic Discourses," 17.

154 Shi Xianpeng and Wu Changbin, "基于海洋命运共同体理念的深海战略新疆域建设" [Construction of a Deep Sea Strategic New Frontier Based on the Concept of

the Maritime Community of a Shared Future], 海洋开发与管理 [*Ocean Development and Management*] 4 (2020): 17–22.

155 "《2021中国的航天》白皮书" [White Paper on China's 2021 Space Program], 中华人民共和国国务院新闻办公室 [*The State Council Information Office of the PRC*], January 28, 2022, www.scio.gov.cn/zfbps/32832/Document/1719689/1719689. htm.

156 Yang Jian, "以人类命运共同体思想引领新疆域的国际治理" [The International Governance of New Frontiers Guided by the Vision of the Community of Shared Future for Mankind], 当代世界 [*Contemporary World*], June 23, 2017, http://cpc.people. com.cn/n1/2017/0623/c191095-29358375.html.

157 Dong Yongzai, "极地安全：国家安全的新疆域" [Polar Security: A New Frontier for National Security], 光明日报 [*Guangming Daily*], April 25, 2021, https://epaper.gmw. cn/gmrb/html/2021-04/25/nw.D110000gmrb_20210425_1-07.htm.

158 Andersson, "The Arctic as a 'Strategic' and 'Important' Chinese Foreign Policy Interest: Exploring the Role of Labels and Hierarchies in China's Arctic Discourses," 17.

159 "《习近平关于网络强国论述摘编》出版发行" [Excerpts on Xi Jinping's Cyber Great Power Analysis Were Published], *Xinhua*, January 21, 2021, www.xinhuanet. com/politics/leaders/2021-01/21/c_1127009956.htm.

160 "海洋局：加快推动我国向极地强国迈进" [State Oceanic Administration: Accelerate China's March Towards Becoming a Polar Great Power], 国家海洋局 [*The State Oceanic Administration*], June 30, 2016, www.gov.cn/xinwen/2016-06/30/content_5087041.htm.

161 "习近平：向海洋进军，加快建设海洋强国," [Xi Jinping: March Toward the Seas and Speed Up Efforts to Build China into a Maritime Great Power], 党建网 [*Party Building Network*], June 8, 2022, www.dangjian.com/shouye/dangjianyaowen/202206/t20220608_6398476.shtml.

162 "《2021中国的航天》白皮书" [White Paper on China's 2021 Space Program], 中华人民共和国国务院新闻办公室 [*The State Council Information Office of the PRC*].

163 Yang, "推动构建人类命运共同体" [Promoting the Building of a Community with a Shared Future for Mankind].

164 Wang, "落实全球安全倡议，守护世界和平安宁" [Implement the Global Security Initiative, Safeguard World Peace and Stability].

165 Yang, "以人类命运共同体思想引领新疆域的国际治理" [The International Governance of New Frontiers Guided by the Vision of the Community of Shared Future for Mankind].

166 Ibid.

167 Deng, "全球公域视角下的极地安全问题与中国的应对" [Analysis of Polar Security Issues from the Perspective of Global Commons and China's Countermeasures], 32.

168 Xiao et al., *The Science of Military Strategy 2020*, 166.

169 "国务院办公厅关于推动国防科技工业军民融合深度发展的意见" [State Council Suggestions on the Promotion of the In-Depth Development of MCF of Defence Science, Technology and Industry], 中华人民共和国中央人民政府 [*The State Council of the PRC*], December 4, 2017, www.gov.cn/zhengce/content/2017-12/04/content_5244373.htm. For an in-depth analysis of the PRC's MCF in the maritime, cyber and space domains see Stone and Wood, *China's Military-Civil Fusion Strategy: A View from Chinese Strategists*, 93–103.

170 Deng, "全球公域视角下的极地安全问题与中国的应对" [Analysis of Polar Security Issues from the Perspective of Global Commons and China's Countermeasures], 37.

171 Yang, "以人类命运共同体思想引领新疆域的国际治理" [The International Governance of New Frontiers Guided by the Vision of the Community of Shared Future for Mankind].

172 Zhang and Liu, "当前国际法跨学科人才培养的新任务新课题 – 基于深海、极地、外空、网络等"战略新疆域"的思考" [New Tasks and Topics for Current

International Law Interdisciplinary Talent Training – Based on Deep Sea, Polar Regions, Outer Space, Cyber Space and Other Strategic New Frontiers].

173 Zhang Zhizhou, "增强中国在国际规则制定中的话语权" [Enhancing China's Discourse Power in International Rule-Making], 人民日报 [*People's Daily*], February 17, 2017, www.scio.gov.cn/zhzc/10/Document/1542461/1542461.htm.

174 Rolland, *China's Vision for a New World Order*, 49.

175 Doshi, *The Long Game: China's Grand Strategy to Displace American Order*, 35–36.

176 Zhao Suisheng, "China's Foreign Policy Making Process: Players and Institutions," in *China and the World*, ed. David Shambaugh (Oxford: Oxford University Press, 2020), 86.

177 David M. Lampton, "Xi Jinping and the National Security Commission: Policy Coordination and Political Power," *Journal of Contemporary China* 24, no. 95 (2015): 773.

178 Hu Weixing, "Xi Jinping's 'Major Country Diplomacy': The Role of Leadership in Foreign Policy Transformation," *Journal of Contemporary China* 28, no. 115 (2021): 13.

179 Jean-Pierre Cabestan, "China's Foreign and Security Policy Institutions and Decision-Making Under Xi Jinping," *The British Journal of Politics and International Relations* 23, no. 2 (2021): 328–329.

180 David Shambaugh, *China Goes Global: The Partial Power* (Oxford: Oxford University Press, 2013), 66.

181 Linda Jakobson and Dean Knox, "New Foreign Policy Actors in China," *SIPRI Policy Paper* 26 (2010): 6.

182 Cabestan, "China's Foreign and Security Policy Institutions and Decision-Making Under Xi Jinping," 327.

183 Jakobson and Knox, "New Foreign Policy Actors in China," 1–51; Shambaugh, *China Goes Global: The Partial Power*, 61–72; Thomas J. Christensen, "More Actors, Less Coordination? New Challenges for the Leaders of a Rising China," in *China's Foreign Policy: Who Makes It, and How Is It Made?* ed. Gilbert Rozman (New York: Palgrave Macmillan, 2013), 23–37.

184 For example, see "Norwegian Deputy Oil Minister to Meet China's CNOOC, as Oslo-Beijing Relations Thaw," *South China Morning Post*, April 6, 2017, www.scmp.com/news/china/economy/article/2085187/norwegian-deputy-oil-minister-meet-chinas-cnooc-oslo-beijing.

185 For an ongoing academic debate on whether or not SOEs are an instrument of the PRC's foreign strategy, see Nie Wenjuan, "China's State-owned Enterprises: Instruments of Its Foreign Strategy?" *Journal of Contemporary China* 31, no. 135 (2022): 383–397.

186 Jakobson and Knox, "New Foreign Policy Actors in China," 24.

187 Silvia Menegazzi, "Chinese Think Tanks and Public Diplomacy in the Xi Jinping Era," *Global Society* 35, no. 3 (2021): 374.

188 "Xi Calls for New Type of Think Tanks," *Xinhua*, October 27, 2014, www.chinadaily.com.cn/china/2014-10/27/content_18810882.htm.

189 Menegazzi, "Chinese Think Tanks and Public Diplomacy in the Xi Jinping Era," 374–375.

190 Nadège Rolland, *Commanding Ideas: Think Tanks as Platforms for Authoritarian Influence* (Washington, DC: National Endowment for Democracy, 2020), 7.

191 For example, see Roderick Kefferpütz, "Big Fish in Small Ponds: China's Subnational Diplomacy in Europe," *MERICS*, November 18, 2021, https://merics.org/en/report/big-fish-small-ponds-chinas-subnational-diplomacy-europe.

192 Flora Yan, "PRC Perspectives on Subnational Diplomacy in China-US Relations," *The Diplomat*, December 16, 2021, https://thediplomat.com/2021/12/prc-perspectives-on-subnational-diplomacy-in-china-us-relations/.

193 Yang Yi, "中国外交决策中的地方政府 – 以广西推动"泛北部湾区域经济合作"为例" [Local Governments in China's Foreign Policy Decision-Making – Guangxi's Promotion of Pan-Beibu Gulf Regional Economic Cooperation as an Example], 理论月刊 [*Theory Monthly*] 5 (2016): 110–116.

194 Chen Xiang and Wei Hong, "一带一路建设视野下的中国地方外交" [China's Local Diplomacy under the Perspective of the Belt and Road Initiative], 国际观察 [*International Review*] 6 (2016): 31–43.

195 Martin Kossa, "China's Arctic Engagement: Domestic Actors and Foreign Policy," *Global Change, Peace & Security* 32, no. 1 (2020): 26.

196 For an in-depth analysis of how the polar regions are handled in the CCP's hierarchy and institutions see: Anne-Marie Brady, *China as a Polar Great Power* (Cambridge: Cambridge University Press, 2017), 114–118.

197 Linda Jakobson and Jingchao Peng, "China's Arctic Aspirations," *SIPRI Policy Paper* 34 (2012): 3–4; Aki Tonami, *Asian Foreign Policy in a Changing Arctic: The Diplomacy of Economy and Science at New Frontiers* (London: Palgrave, 2016), 28–33; Brady, *China as a Polar Great Power*, 118–122.

198 "自然资源部职能配置、内设机构和人员编制规定" [Functions, Internal Organization and Staffing Provisions of the Ministry of Natural Resources], 中华人民共和国自然资源部 [*The Ministry of Natural Resources of the PRC*], September 11, 2018, www.mnr.gov.cn/jg/sdfa/201809/t20180912_2188298.html.

199 "单位职责" [Unit Responsibilities], 国家海洋局极地考察办公室 [*Chinese Arctic and Antarctic Administration*], n.d., http://chinare.mnr.gov.cn/catalog/organization.

200 Linda Jakobson and Seong-Hyon Lee, *The North East Asian States' Interests in the Arctic and Possible Cooperation with the Kingdom of Denmark* (Stockholm: Stockholm Peace Research Institute, 2013), 7.

201 "新版《北极航行指南（东北航道）》正式出版发行" [A New Version of the Navigation Guidelines for the Arctic (Northeast Passage) Were Published], 中国水运报 [*China Water Transport*], July 28, 2022, www.msa.gov.cn/html/xxgk/hsyw/20220728/46BC9AE6-0371-4121-BBC4-72598FCD4343.html.

202 "极地水域操作手册编写指南2017" [Polar Waters Operations Manual Preparation Guide 2017], 中国船级社 [*China Classification Society*], December 27, 2016, www.ccs.org.cn/ccswz/articleDetail?id=201900001000008151.

203 "职能配置" [Allocation of Functions and Responsibilities], 中华人民共和国农业农村部 [*The Ministry of Agriculture and Rural Affairs of the PRC*], April 10, 2019, www.moa.gov.cn/jg/bjs/201712/t20171217_5986506.htm.

204 "农业农村部关于促进"十四五"远洋渔业高质量发展的意见" [Opinions of the Ministry of Agriculture and Rural Affairs on the Promotion the High-quality Development of Distant Ocean Fisheries During the 14th Five-Year Plan], 中华人民共和国农业农村部 [*The Ministry of Agriculture and Rural Affairs of the PRC*], February 14, 2022, www.moa.gov.cn/govpublic/YYJ/202202/t20220215_6388748.htm.

205 "科学技术部职能配置、内设机构和人员编制规定" [Functions, Internal Organization and Staffing Provisions of the Ministry of Science and Technology], 中华人民共和国科学技术部 [*The Ministry of Science and Technology of the PRC*], n.d., www.most.gov.cn/zzjg/kjbzn/201907/t20190709_147572.html.

206 "中俄高技术和创新工作组第十三次会议在线举行" [The 13th Meeting of the Sino-Russian High-Tech and Innovation Working Group Was Held Online], 中华人民共和国科学技术部 [*The Ministry of Science and Technology of the PRC*], June 15, 2022, www.most.gov.cn/kjbgz/202206/t20220615_181151.html.

207 "21世纪中心主办金砖国家"海洋与极地科学"专题领域工作组第四届会议" [The 21st Century Centre Hosted the Fourth Meeting of the BRICS Working Group on Marine and Polar Sciences], 中华人民共和国科学技术部 [*The Ministry of Science and Technology of the PRC*], August 6, 2021, www.most.gov.cn/kjbgz/202108/t20210806_176297.html.

208 "工业和信息化部主要职责内设机构和人员编制规定印发" [The Main Responsibilities and Regulations on the Internal Structure and Staffing of the Ministry of Industry and Information Technology Were Published], 中华人民共和国工业和信息化部 [*The Ministry of Industry and Information Technology of the PRC*], September 16, 2015, www.miit.gov.cn/gyhxxhb/jgzz/art/2020/art_764adf9bbab147c39c934519f8e1103b.html.

209 "中芬商谈在北极海底铺光缆 中国或获进入北极新入口" [China and Finland Are Talking about Laying a Fibre-Optic Cable on the Arctic Ocean Floor – China May Get a New Entry into the Arctic], 参考消息 [*Reference News*], December 19, 2017, www.chinanews.com/gn/2017/12-19/8403497.shtml.

210 "教育部直属高等学校" [Institutions of Higher Learning Directly under the Ministry of Education], 中华人民共和国教育部 [*The Ministry of Education of the PRC*], n.d., www.moe.gov.cn/jyb_zzjg/moe_347/.
211 "海洋生态环境司" [Department of Marine Ecological Environment], 中华人民共和国生态环境部 [*The Ministry of Ecology and Environment of the PRC*], n.d., www.mee.gov.cn/zjhb/bjg/hys/.
212 For example, see "第二届中俄能源商务论坛在圣彼得堡举办" [The Second China-Russia Energy Business Forum Was Held in St. Petersburg], 国家能源局 [*National Energy Administration*], June 14, 2019, www.nea.gov.cn/2019-06/14/c_138143530.htm.
213 For example, see Mary Ilyushina, "Russia Accuses Leading Arctic Researcher of Spying for China," *CNN*, June 17, 2020, https://edition.cnn.com/2020/06/17/europe/russia-china-spying-allegation-intl/index.html.
214 Brady, *China as a Polar Great Power*, 119.
215 Kossa, "China's Arctic Engagement: Domestic Actors and Foreign Policy," 35.

3 The Arctic in Chinese Science and Research

Science, research, and innovation play a pivotal role in the PRC's national rejuvenation endeavours, serving as catalysts of socio-economic development and military modernisation. The CCP is allocating increasingly substantial resources to research and large-scale scientific projects, aiming to position China as a leading science and technology power. The PRC's 14th Five-Year Development Plan, spanning from 2021 to 2025, emphasises the enhancement of China's national strategic science and technology capabilities. This entails prioritising cutting-edge fields such as next-generation AI, quantum information, integrated circuits, neuroscience, biotechnology, and clinical medicine. Furthermore, the plan maintains a sustained focus on researching strategic new frontiers, including the polar regions. Specifically, the development plan mandates the improvement of Chinese polar multidimensional observation platforms and the implementation of Phase 2 of the PRC's national-level research project, Xuelong Exploring the Polar Regions (雪龙 探极).[1] According to Chinese officials, the underlying objective of this project, initially launched during the 13th Five-Year Development Plan (2016–2020), is to enhance China's operational capabilities for polar scientific observations. This involves expanding the PRC's polar scientific infrastructure, observation networks, and logistical capabilities (including new research stations, vessels, and vehicles), constructing communication systems between stations and ships to facilitate effective data transmission, and developing application systems, such as weather and sea ice prediction models.[2]

Consequently, it is expected that China, in its New Era of development, will continue deepening its scientific engagement with the Arctic region. This stems from several factors. First, the Arctic is seen as a region that holds considerable scientific value in Chinese official and scientific discourses. Chinese officials have noted that the Arctic has great utility "as an indicator of global climate change and a laboratory for global scientific research."[3] Moreover, Chinese scientists highlight the region's influence on Northern Hemisphere weather patterns, the impact of Arctic Ocean water circulation on the North Pacific,[4] and its suitability for studying the upper atmosphere, geomagnetism, and outer space phenomena.[5]

This chapter is divided into two main sections. The first section outlines Chinese efforts to establish a multidimensional Arctic research and monitoring system, relying on a network of domestic research organisations, in situ research platforms

DOI: 10.4324/9781003295112-3

(such as science stations), and remote sensing capabilities (such as polar-observing satellites). The second section explores the implications of these efforts and discusses their significance to the Party-state in terms of climate research and ecological security, international scientific cooperation, and the process of Chinese Arctic identity formation.

Towards a Multidimensional Arctic Research and Monitoring System

In recent years, Chinese science organisations have made significant strides in deploying new technologies and establishing advanced capabilities for studying the Arctic region. These include the construction of research stations, icebreakers, meteorological platforms, buoys, unmanned underwater vehicles (UUVs), as well as the development of remote sensing infrastructure such as polar-orbiting satellites, unmanned aerial vehicles, and satellite ground-receiving stations. Concurrently, the PRC has intensified its research efforts in Arctic-related social sciences to understand the impact of the region's opening on its own socio-economic development and international relations. The ultimate objective of the PRC is to construct an integrated Arctic scientific observation and monitoring network that encompasses space, atmosphere, land, ice, and underwater dimensions.[6] These efforts are led by a growing network of Chinese research organisations and academic institutions.

Chinese Polar Research Organisations

The Polar Research Institute of China (PRIC 中国极地研究中心), headquartered in Shanghai and directly managed by the MNR, serves as the operational hub for Chinese Arctic and Antarctic science, research, and expeditions. In addition to conducting on-site research in the polar regions, the Institute is responsible for tasks, such as the construction, operation, and management of China's polar research platforms, including field stations, as well as research expedition support systems comprising research ships, aircraft, and other equipment. Moreover, the Institute oversees the management of the PRC's polar archives, data, and samples, while actively engaging in international cooperation, exchanges, and the popularisation of polar science.[7]

PRIC leads China's research endeavours in polar glaciology and oceanography, with a focus on their connections to global climatic changes and broader atmospheric processes. It also conducts research on polar biology and ecology, including the study of polar marine ecosystems and their responses to climate change. Additionally, the Institute engages in polar space physics and astronomy to support services related to space weather and debris monitoring.[8] Through its Division of Polar Policy Studies, PRIC also conducts research on strategic issues related to the polar regions, encompassing legal, political, economic, and social aspects. This research aims to provide policy and strategy advice to decision-makers in the PRC.[9] The Institute is also committed to the development of observation technology and equipment systems suited for the unique polar environment. Its Key Laboratory of Polar Science strives to be at the forefront of these endeavours, advancing the

MNR's mission of exploration and utilisation of the polar regions while aiming to "seize the commanding heights of international competition in polar science and technology."[10] PRIC oversees the publication of two journals: the *Chinese Journal of Polar Research* (极地研究) in Chinese since 1988 and *Advances in Polar Science* in English since 1990.[11] Both journals are published quarterly, with open access, and have a multidisciplinary focus. With regards to the Arctic, the majority of articles in these journals focus on natural science research. However, there has been an increase in social science–related articles in the past decade. These include topics such as Arctic international law, shipping, and governance, reflecting the growing significance of such themes in China's Arctic approach and national strategy.

PRIC is also actively involved in international scientific cooperation with foreign partners. It represents the PRC at various Arctic-related gatherings, including the Arctic Frontiers in Norway, hosts foreign scientific and political delegations, establishes bilateral science cooperation agreements with foreign counterparts, engages regional and international partners through international scientific networks, such as the Asian Forum for Polar Sciences and the China-Nordic Arctic Research Centre, and participates in long-term international scientific projects such as the MOSAiC expedition. Given this broad international exposure and extensive scope of Arctic (and Antarctic) research interests, PRIC serves as a nexus between scientific research, the international Arctic scientific community, and the Chinese Party-state. Effectively, PRIC functions as an organisation that gathers and processes data, providing relevant information to the Chinese political system to assist decision-makers in policy deliberation.

Additionally, several universities possess significant expertise in Arctic-related research fields, spanning from international law to remote sensing and engineering.[12] These universities include the following:

- Dalian Maritime University, which operates the Arctic Maritime Affairs Research Centre, focusing on Arctic navigation and traffic safety, resource development and environmental protection, as well as international law and policy;[13]
- Harbin Engineering University (HEU), which houses the Polar Science and Engineering Research Centre, specialising in technological areas such as polar acoustics, communication, navigation, and materials.[14] HEU also manages the Arctic Blue Economy Research Centre, promoting cooperation between China and Russia in Arctic sustainable development, science and technology, and education;[15]
- Harbin Institute of Technology (HIT), which oversees the Polar Academy, researches the Arctic environment and ecosystem, polar space environment, engineering, transportation, architecture, and biology;[16]
- Liaocheng University, with its Arctic Studies Centre dedicated to exploring cultural and ecological issues, emphasising archaeological, anthropological, historical, and biological perspectives;[17]
- Ocean University of China in Qingdao houses the Polar Research Centre, which focuses on various aspects including polar geopolitics and strategy, governance and legal issues, as well as development and resource utilisation;[18]

- Shanghai Jiaotong University, which operates the Centre for Polar and Deep Ocean Development, with a specific focus on polar shipping, environmental protection, and the development of strategic resources;[19]
- Sun Yat-sen University in Zhuhai, with its School of Geospatial Engineering and Science, has a strong focus on polar and ocean remote sensing;[20]
- Tongji University in Shanghai houses the Centre for Polar and Oceanic Studies, specialising in Arctic politics, law, security, economy, and China's polar strategy and policy;[21]
- Wuhan University, which manages the China Institute of Boundary and Ocean Studies, and conducts research on polar strategy and interests.[22]

In 2018, 25 Chinese universities collaborated to establish the Joint Centre for Polar Research of Chinese Universities (中国高校极地联合研究中心) with the goal of conducting joint studies on Arctic and Antarctic environmental change, shipping, resource development, and the economy.[23] Brady observes that such clustering of universities indicates a substantial focus of the Party-state on Arctic-related matters, as Chinese researchers tend to assemble around topics of high policy interest.[24]

In addition to the universities, the Shanghai Institutes for International Studies, one of the most influential foreign policy think tanks in China under the management of the MFA, is engaged in projects related to Arctic affairs. The think tank operates the Centre for Marine and Polar Studies, which focuses on maritime and Arctic affairs, Chinese maritime development strategy, Chinese Arctic policies, and cooperation between China and Arctic countries.[25] The China Meteorological Administration (CMA), under the PRC State Council, supports polar meteorological observations and research projects through its subordinate units like the Chinese Academy of Meteorological Sciences. It also installs automated weather stations in the polar regions and engages in international polar cooperation.[26] The Chinese Academy of Sciences (CAS), including institutes such as the Institute of Oceanology, the Research Centre for Eco-Environmental Sciences, and the Institute of Atmospheric Physics also engage in Arctic scientific research.[27] The China Geological Survey (CGS), under the MNR, operates the Research Center of Polar Geosciences, aiming to develop a polar terrestrial and deep-sea geological observation system.[28] In addition, the First, Second, and Third Institutes of Oceanography, all under the MNR, conduct polar-related research and participate in China's Arctic scientific expeditions.[29]

Chinese Arctic Field Monitoring Infrastructure

The Arctic Yellow River Earth System National Field Science Observation and Research Station (北极黄河地球系统国家野外科学观测研究站)

The Yellow River Station is the PRC's inaugural official Arctic research station, established in July 2004 in Ny-Ålesund, in the Spitsbergen Archipelago of Norway. Ny-Ålesund is home to the world's northernmost science community, equipped with state-of-the-art research facilities.[30] The research station consists of a two-story building covering approximately 500 m², housing laboratories, offices, a dormitory, storage facilities, and an observation platform on the rooftop. It can accommodate up to 20 people at a time.[31] This is comparable in size to the UK's Arctic station[32] but twice the size of India's Arctic station, Himadri.[33] Since its establishment, the PRC has utilised the research station for annual science expeditions both during the summer and winter seasons. Research activities at the Yellow River Station primarily focus on glaciology, climate studies, environmental monitoring, upper atmospheric physics, space environment and physics, geology, microbiology, marine biology, meteorology, and sea ice formations using unmanned aerial technology.[34]

The China-Iceland Arctic Science Observatory (CIAO 中冰北极科学考察站)

The establishment of the CIAO project in Kárhóll, northern Iceland, was the outcome of Sino-Icelandic scientific collaborations that began expanding in 2012 with the signing of an MoU on research cooperation between PRIC and the Icelandic Centre for Research (RANNÍS).[35] The Observatory, inaugurated in 2018, primarily supports multidisciplinary observation and research on auroras, atmospheric science, meteorology, and satellite remote sensing.[36] Upper atmospheric observations in Iceland contribute strategically to the development of China's space program by filling knowledge gaps in these areas, as auroras serve as important indicators of space weather that can affect satellite operations, space stations, and human space activities.[37] The Observatory is being adapted to accommodate research in other scientific fields, including oceanography, geophysics, and biology.[38] The cost of the construction, estimated at around USD 2.4 million, was financed by the Chinese Party-state through PRIC.[39] In addition to Chinese and Icelandic scientists, CIAO is also expected to host visiting international researchers, showcasing the positive research cooperation relations between Iceland and China, which have developed alongside their cordial economic and political ties.

Chinese Arctic specialists have suggested that considering the PRC's expanding Arctic scientific interests, the possibility of establishing another research station in the region, potentially in Greenland or the Russian Arctic, should not be overlooked.[40] This was reaffirmed in 2016 when an official PRC delegation visited Greenland, leading to the signing of an MoU on scientific cooperation with the Greenlandic Ministry of Education, Culture, Research, and Church. The agreement emphasised that "regarding the joint construction of a station, the parties will maintain active communication and collaboration."[41] Subsequently, in 2017, during the Arctic Circle Assembly in Iceland, PRIC announced China's desire to establish a research station in Greenland, proposing two potential locations: one in the far north of Greenland and the other in the southwest of the island.[42] However, since then, potentially due to heightened international scrutiny of its Arctic activities, the PRC has not provided any further information regarding the development of this proposed research station.

Research Icebreakers: Xuelong and Xuelong 2

Another platform utilised by the PRC for researching, observing, and monitoring the Arctic region and its unique environment is research vessels. Currently, the

PRC operates two icebreakers, both of which have undertaken research expeditions in the region. Xuelong, acquired from Ukraine in 1993, was commissioned in 1994 after undergoing retrofits to replace the R/V Jidi. It is one of the world's largest non-nuclear icebreakers, measuring 167 m in length with a total displacement of 21,025 tons. Capable of breaking through 1.2 m of ice at a speed of 1.5 knots, the vessel can accommodate approximately 120 people. It is equipped with navigational, positioning, and automatic piloting systems, a platform for two helicopters, a hangar, as well as laboratories for marine physics, marine chemistry, biology, and a data processing centre.[43] Additionally, it provides crucial logistical support for Chinese Antarctic expeditions, facilitating the transportation of cargo to and from Chinese Antarctic bases. The research vessel has undergone two extensive upgrades, in 2007 and 2013.[44]

Recognising the growing demands of its polar scientific activities, the PRC State Council announced in 2009 the need for new research vessels with icebreaking capabilities.[45] Consequently, in 2012, the PRC selected the Finnish company Aker Arctic to perform the conceptual and basic design of its new polar research vessel. The No. 708 Research Institute of the China State Shipbuilding Corporation (CSSC) provided technical support for this project and played a leading role in optimising the new ship's scientific, intelligent, and open water performance capabilities.[46] CSSC is the PRC's primary corporation for the research, design, production, and testing of naval vessels, weapons, and equipment, including aircraft carriers and nuclear submarines.[47] Construction on the new icebreaker, named Xuelong 2, officially commenced in Shanghai at the Jiangnan shipyards in 2016, and the ship was launched in September 2018. With a length of 122 m, the vessel is equipped with laboratories, several cranes, a moon pool with a scientific hangar, aviation facilities, and a landing platform for helicopters. It comfortably accommodates up to 90 scientists and crew members. The Xuelong 2 can continuously break up to 1.5 m thick ice in both forward and backward directions, making it highly manoeuvrable in complex sea ice conditions.[48] Xuelong 2 is capable of conducting comprehensive surveys of the polar oceans, sea ice, and the atmosphere. It can also perform multidomain observations and sampling of the marine environment related to climate change, undertake geophysical and seismic exploration, and survey seabed topography and biological resources in polar areas.[49] According to Wu Gang, the chief designer of the new research icebreaker, Xuelong 2 has provided the PRC with crucial technical insights into advanced polar technology, including the two-directional propulsion system, Polar Class (PC) 3 icebreaker structure design technology, low-temperature and cold design protection technology, and welding technology of icebreaker construction.[50] This new generation of polar research icebreakers marks a qualitative advancement in China's polar research capabilities. With the enhanced capabilities of Xuelong 2, such as the moon pool and two-directional propulsion system, Chinese scientists will be able to expand the scope of their scientific expeditions with advanced options to operate under unique and diverse polar conditions.[51]

As of June 2022, Chinese research vessels have completed 12 Arctic scientific expeditions: nine by Xuelong, two by Xuelong 2, and one by the oceanographic research vessel Xiang Yang Hong 1. These expeditions typically occur between July and September and last approximately 70 to 90 days. The initial two voyages into the Arctic region, launched in 1999 and 2003, were relatively irregular. However, since 2008, the expeditions have become institutionalised to some degree, with a consistent number of participating researchers, an expanding scope of studies, systematised research tasks, and a regular interval of one Arctic scientific expedition every two years.[52] In 2017, the SOA announced that China would increase the frequency and intensity of its Arctic research expeditions to one per year. This decision was driven by the fact that China, as a near-Arctic state, is heavily influenced by environmental changes in the Arctic, which impacts its environment and socio-economic developments. The PRC sought more information and knowledge to develop a better understanding of these changes and their effects, as stated by Lin Shanqing, the deputy director of the SOA at the time.[53]

Chinese Arctic scientific expeditions primarily take place in the Bering and the Chukchi Seas, as well as the wider Canada Basin in the Arctic Ocean. Occasionally, they venture beyond this area to circumnavigate the Arctic, demonstrating the growing capabilities and confidence of the Chinese Arctic research program.[54] Each expedition typically involves over 100 participants, including researchers from various Party-state institutions such as PRIC, the First, Second, and Third Institute of Oceanography, the National Marine Environmental Forecasting Centre, and Chinese universities. In addition to researchers, constituting about half of the expedition members, there were 30–40 crew members, CCTV and Xinhua reporters, and other supporting staff.[55] Foreign scientists are also invited to participate in these scientific expeditions, reflecting the Chinese Party-state's efforts to showcase its inclusive Arctic scientific approach. The scientific tasks carried out during Arctic voyages are multidimensional in nature, including comprehensive surveys of the marine environment, sea ice, and the atmosphere, as well as biological, geological, and geophysical observations.

Meteorological Stations

Since 2012, Chinese scientists have been installing automatic weather stations in the Arctic to enhance their regional monitoring capabilities.[56] These stations are predominantly deployed on drifting sea ice in the Arctic Ocean but have also been established on Svalbard.[57] Standing at around 4 m tall, these stations collect meteorological data, including wind speed and direction, temperature, humidity, atmospheric pressure, and radiation. Equipped with advanced ultralow temperature batteries and solar panels, some stations can provide continuous services for months, transmitting real-time data back to the PRC via satellite link.[58] These meteorological observations not only contribute to a better understanding of Arctic environmental changes but also support the optimisation of Arctic atmospheric and sea ice numerical forecasting models. These models, as noted in a report, provide additional information for the strategic use of Arctic sea routes and the expansion of the Polar Silk Road.[59] Polar meteorology has been identified as an important focus area in the PRC's meteorological sciences. The recently unveiled 2021–2035

China Meteorological Science and Technology Development Plan directs Chinese scientists to research climate and weather changes along the Arctic shipping routes, polar air pollution, polar atmospheric variations, and their impact on weather and climate in East Asia, as well as the influence of melting Arctic sea ice and atmospheric circulation on mid-latitude extreme weather and climate events.[60]

European Incoherent Scatter Scientific Association (EISCAT)

The PRC's interest in atmospheric studies is further demonstrated through its participation in EISCAT, an international scientific organisation that conducts ionospheric and atmospheric measurements using radar. EISCAT operates in three Arctic states, with radar antennas located in Kiruna, Sweden; Sodankylä, Finland; Tromsø, Norway; and in Longyearbyen, Svalbard, also in Norway – all situated well north of the Arctic Circle.[61] The China Research Institute of Radio Wave Propagation (also known as China Electronics Technology Group Corporation No. 22 Research Institute) represents the Party-state within this association. It is the only national-level institute in the PRC specialising in radio wave propagation observation and research, with its outputs finding significant applications in radar technology, communication, navigation, remote sensing, space flight, and radio interference.[62] In 2014, the PRC sought to construct a new large radar antenna in Longyearbyen, Svalbard, to augment the existing EISCAT installations. However, due to security reasons, the Norwegian Government denied permission for China to build such a facility, stating that "the goal of the installation is to conduct research on the upper reaches of the atmosphere, but the technology also has other uses," and that "after an overall evaluation, we don't want such an antenna on Svalbard."[63]

Buoys

To obtain Arctic oceanographic measurements, Chinese scientists use buoys as versatile platforms for observing the ocean and sea ice. During previous Chinese Arctic expeditions, buoys were deployed in ice-free areas of the Norwegian Sea to study air-sea interactions,[64] as well as the North Pacific Ocean to collect meteorological data such as temperature, air pressure, wind speed, sea surface temperature, and ocean salinity measurements.[65] Chinese researchers have also placed similar ice-buoy systems in ice-covered regions of the Arctic Ocean to investigate the interactions between the upper layers of the ocean and sea ice, as well as the processes driving Arctic sea ice changes.[66] For instance, during the PRC's seventh Arctic scientific expedition in 2016, dozens of ice-based buoys were deployed in the Canada Basin. These buoys drifted on the sea ice, continuously monitoring it and gathering data on sea ice movement, extent, and various thermodynamic processes, such as snow accumulation and melting on the sea ice surface.[67] Chinese universities, in particular, are actively involved in the development of these systems. The Ocean University of China independently constructed a buoy system called the Drift-Towing Ocean Profiler (D-TOP), which enables year-round real-time monitoring of the Arctic marine environment.[68] The Taiyuan University of

Technology has built and deployed various sea ice multiparameter buoys, allowing for the study of the acoustic environment beneath Arctic ice and long-term observation of Arctic sea-ice-air interactions.[69]

Furthermore, Chinese scientists have been integrating the aforementioned meteorological and oceanographic systems and technologies. In 2018, during the ninth Arctic scientific expedition, they installed their first unmanned ice station, aiming to achieve a more comprehensive monitoring system of the Arctic environment.[70] This system, known as the Arctic Sea-Ice-Air Unmanned Ice Station Observation System (北极海 – 冰 – 气无人冰站观测系统), consists of four components: a meteorological tower, a main buoy on ice (for sea ice observation), a sub-buoy (for upper ocean layer observation), and a towed ocean profile observation instrument. The system is used to observe the effects of atmospheric processes and the upper layers of the ocean on the growth and melting of Arctic sea ice. Equipped with multiple sensors, including a dissolved oxygen sensor, temperature and air pressure sensor, humidity sensor, and positioning sensor to track the movement of Arctic sea ice.[71] Lei Ruibo from PRIC notes that since these unmanned ice stations can remain in the Arctic for several months, these systems enable the PRC to obtain important data throughout the year, even when Chinese icebreakers are not in the region. This expands their ability to monitor Arctic environmental changes and improves the accuracy of predicting Arctic sea ice variations.[72]

Unmanned Underwater Vehicles

To survey the under-sea environment, Chinese scientists have been deploying numerous autonomous and remotely controlled unmanned underwater vehicles in the Arctic Ocean and adjacent seas since 2008. These UUVs are capable of carrying a wide array of scientific instruments.[73] The Shenyang Institute of Automation under CAS has been at the forefront of developing and testing these UUVs. The Arctic ARV series of vehicles, launched during the 2008, 2010, and 2014 Arctic scientific expeditions, were used to collect various scientific observation data, including water temperature, salinity, depth, subglacial light transmission irradiance, ice bottom shape, and sea ice thickness. These UUVs greatly complement the environmental monitoring capabilities of Chinese scientists.[74] In 2018, during the PRC's ninth Arctic scientific expedition, Chinese scientists successfully launched and recovered an underwater glider, which operated for 45 days in the Bering Sea, acquiring continuous temperature and water salinity data.[75] More recently, during the 12th Arctic expedition, Chinese researchers tested a new, more advanced generation of UUVs, the Tansuo (探索) 4500. These UUVs were used to observe the thickness and movement of sea ice, as well as various ocean parameters such as depth, temperature, salinity, and ocean topography.[76] Apart from CAS, other renowned research institutions in the PRC are also developing polar UUV technology. For instance, according to Cheng Xiao et al., Harbin Engineering University, under the Ministry of Industry and Information Technology, has developed a prototype of a polar autonomous underwater vehicle capable of reaching depths of 1,000 m, collecting hydrological data, and observing submarine topography and sea ice formation.[77]

Chinese Arctic Remote Sensing Capabilities

The use of space-based assets, such as polar-orbiting satellites, has become an increasingly important aspect of the PRC's capabilities in observing, monitoring, and understanding the unique Arctic environment. According to Chinese experts, the PRC currently has over 20 remote sensing satellites in orbit that can, to some extent, observe the polar regions and gather information on sea ice, land formation, and the marine environment.[78] These include the Gaofen (高分), Ziyuan (资源), Fengyun (风云), and Haiyang (海洋) series of high-resolution satellites. The Gaofen satellites possess multi-observational capabilities with high spatial, spectral, and temporal resolution. Advanced versions like the Gaofen 3 satellite are equipped with synthetic aperture radar, enabling observation of the Earth's surface regardless of weather conditions or time of day. The Gaofen 3 can penetrate clouds, surface vegetation, ice, and snow.[79] This technology accurately maps and classifies floating sea ice fragments in the polar oceans during the summer melting season. According to Cheng Xiao et al. from Sun Yat-sen University, Chinese research icebreakers and merchant ships have used data from these satellites for navigation in the polar oceans.[80] The Ziyuan series comprise remote sensing satellites that the PRC employs to acquire high-resolution images for surveying Earth resources, disaster management, and ecological and land use monitoring.[81] The Ziyuan 3 series also successfully monitored polar ice sheet surface elevation, shape, and movement.[82] The Fengyun series comprises the PRC's meteorological satellites. Specifically, the Fengyun 3 series is designed to collect multidimensional atmospheric and surface characteristics data and provide weather information worldwide for specialised activities, including aviation, marine activities, and national defence.[83] The Micro-Wave Radiation Imager installed on these satellites complements the PRC's capabilities in polar ice and snow environment monitoring.[84] The Haiyang series of high-resolution satellites are employed by the PRC to monitor the marine environment and gather data on sea surface wind, wave height, and temperature, among others. They are also capable of observing sea ice coverage, concentration, and type.[85] For instance, observational data obtained from the Haiyang satellites supported the completion of scientific tasks during the 10th Chinese Arctic expedition.[86]

While the previous examples demonstrate how Chinese scientists use data from these satellites to study the Arctic region, it is important to note that these satellite series are not specialised polar observing satellites. Therefore, in September 2019, through collaboration between Beijing Normal University, Sun Yat-sen University, and the Shenzhen Aerospace Dongfanghong HIT Satellite company, the PRC successfully launched its first polar observing microsatellite, the Ice Pathfinder (code BNU-1). Weighing only 16 kg, BNU-1 carries optical instruments for panchromatic and multispectral imaging, providing a five-day revisit period of the polar regions up to 85° latitude. This microsatellite is convenient for environmental monitoring in the polar regions.[87] It also features an onboard Automatic Identification System (AIS) receiver, which aids in planning navigation routes for ships in polar regions.[88] Sun Yat-sen University is also developing a more advanced polar observing satellite, set to launch in the early 2020s. Equipped with synthetic aperture radar, this experimental satellite's specific task is to monitor Arctic sea ice along shipping routes in the Arctic Ocean and provide a two-day revisit period, significantly enhancing Chinese regional monitoring capabilities.[89] According to Chinese reports, the ultimate goal is to establish a constellation of 24 polar observing satellites (with BNU-1 being the first) to achieve nearly uninterrupted coverage of the polar regions.[90]

To maximise the effective use of satellite operations and develop practical applications, the PRC requires a broad network of domestic and international ground-receiving satellite stations. China's domestic ground-receiving stations can receive real-time satellite data covering its entire territory, coastal areas, and 70% of the land area in Asia. However, when it comes to overseas data, such as information from the Arctic region stored on Chinese satellites, this can only be transmitted directly to a domestic ground-receiving station when the satellite transits the country. As a result, the obtained data can be several hours, or even days, old.[91] It is therefore imperative for the PRC to establish dedicated ground-receiving stations in the Arctic region, as these would enable faster transmission of acquired data to relevant institutions in the PRC. In this context, the PRC successfully built the Arctic Receiving Station of the China Remote Sensing Satellite Ground Station, in Kiruna, Sweden, approximately 200 km north of the Arctic Circle. Operated by the Institute of Remote Sensing and Digital Earth of the CAS,[92] this satellite ground station, inaugurated on December 15, 2016, is the PRC's first fully owned ground-receiving station constructed overseas. It is a modular unmanned system capable of receiving all-weather, all-time, and multi-resolution satellite data. Its strategic location significantly enhances the transmission efficiency of Chinese satellite data and improves the PRC's capability to access global remote sensing data.[93]

Additionally, reports indicate that CAS expressed interest in cooperating with Finland in satellite data sharing. In 2018, CAS and the Finnish Meteorological Institute agreed to establish the Joint Research Centre for Arctic Space Observation and Information Services (北极空间观测和信息服务联合研究中心) in Sodankylä, Lapland. The Centre, which was in operation for a period of three years until 2021, focused on polar information sharing, satellite data download and capacity development, scientific research, ground experiments, and personnel exchanges.[94] CAS has also engaged in cooperation with Russia in the field of Arctic remote sensing. For instance, in 2019, CAS successfully obtained satellite remote sensing data in near real-time, covering the Bering Strait and parts of Arctic shipping routes, from the Russian Magadan Ground Station. This data was crucial for conducting rapid monitoring of sea ice and its distribution along these waterways.[95]

In addition to space-based remote sensing and on-site observations, Chinese scientists use aerial vehicles to conduct observations of the Arctic region. PRC researchers primarily from Beijing Normal University and Sun Yat-sen University have been developing the Jiying (极鹰) series of unmanned aerial vehicles (UAVs) for China's polar scientific expeditions.[96] These UAVs offer a more cost-effective and accessible alternative to space-based assets, and they can also compensate for the limitations of spatial resolution in remote sensing observations.[97] In 2017, for example, Chinese researchers used the Jiying series to acquire high-resolution remote sensing images in the Disco Bay area of Greenland.[98]

Chinese Arctic Social Science Research

Since 2008, following the Russian flag-planting stunt at the North Pole in the Arctic Ocean and the subsequent news of extensive reductions in Arctic sea ice, along with the publication of the 2008 USGS report on undiscovered oil and gas resources in the Arctic, the PRC started paying significantly greater attention to the changes occurring in the Arctic region and their impact on China's socio-economic development. Numerous Chinese research institutes initiated projects to study these impacts. For instance, in 2013, PRIC conducted research on the economic, legal, and international aspects of utilising Arctic sea routes.[99] Furthermore, in 2015, the Strategic Assessment of Interests of Polar States Project was completed under the State Oceanic Administration. This project consisted of five sub-projects: (1) Polar Geopolitical Research, (2) Polar Resources Strategic Research, (3) Strategic Research on the Development of Polar Science and Technology, (4) Research on the Polar Legal System, and (5) Research on Polar States' Strategies. Each project extensively focused on the Arctic region and urged China, among other things, to engage in the polar environment and resource utilisation to enhance its influence in international polar affairs.[100] The importance of Arctic social science research was underscored in 2015 by then Director of PRIC, Yang Huigen, who highlighted the need for a sustained focus on both natural and social science research as a foundation for China's participation in polar governance.[101]

Chinese preferences for potential future developments in Arctic affairs can be reflected in projects funded by the National Social Sciences Fund of China, considered the most prestigious state research funding agency for social sciences in China.[102] Since Xi's announcement in 2017 that China was entering a New Era, the National Social Science Fund has awarded funding for projects examining various aspects, such as the joint Sino-Russian construction of the Polar Silk Road, Sino-Russian Arctic cooperation, security risk management on Arctic shipping routes, the PRC's diplomatic strategy and the establishment of an Arctic Community of Shared Future, and China's governance of Arctic science and technology.[103]

Challenges of the PRC's Polar Scientific Research

The gradual expansion of Chinese scientific interests in the Arctic has accentuated some of the existing issues in China's polar sciences. Increased interest in the region has led to more actors within the Party-state science system conducting Arctic-related research and participating in Arctic expeditions, which can result in project overlaps and inefficient resource allocation.[104] This concern was raised by Wu Jun, a senior official from the CAA, who emphasised China's weak polar management capabilities and the absence of polar legislation.[105] To address these issues, the SOA implemented the *Administrative Licensing Regulations for Arctic Research Activities* in 2017. These regulations aim to establish a unified governance model for Arctic research and prevent resource fragmentation. According to these guidelines, any individual or company seeking to engage in research activities in the Arctic region must submit a formal application to the PRC State Council and await formal approval.[106]

Another challenge is the perceived dependence of the PRC on foreign polar technology. Yang Huigen, the former Director of PRIC, acknowledged that while China's polar science has made significant progress in the past 30 years, some of its polar equipment and scientific research instruments still rely on foreign imports, which he called the "short-leg" of China's polar research.[107] To address these technological challenges, Beijing established The Polar Science and Technology Commission (极地科学技术委员会) in 2017, a consultative body consisting of over 40 specialists from the CAS, MOST, and other government departments, tasked with advising on the development of China's polar science and technology.[108] The commission's leadership emphasised the need for the accelerated development of field detection technologies and equipment adapted to the unique polar environment.[109] This sustained focus on domestic polar technology development was also reiterated in China's official Arctic White Paper and aligns with the broader efforts of the CCP to achieve technological self-reliance. In this regard, China is making progress. For example, during the 12th Arctic scientific expedition in 2021, the Second Institute of Oceanography led a research project that focused on conducting complex geophysical surveys of the seafloor in the central Arctic Ocean using domestically developed instruments.[110] China's responses to these challenges demonstrate the importance the Party-state ascribes to its goal of becoming a polar great power.

The Importance of Arctic Scientific Research to the PRC

Since the commencement of its engagement in the Arctic in the late 1990s, China's pursuit of Arctic-related science and research activities has advanced three areas crucial to its security and interests. First, Chinese Arctic scientific research, particularly in understanding the environmental changes occurring in the region and their potential impact on the Chinese mainland, holds important implications for maintaining Chinese ecological security (生态安全). Ecological security, as outlined in Chapter 2, is one of the 16 types of security emphasised in China's Comprehensive National Security Outlook under Xi's leadership. Maintaining ecological security, from the Chinese perspective, is directly related to people's well-being, sustainable economic development, long-term social stability, and the ability to address major internal and external ecological challenges.[111] Both Chinese and international scientists suggest that China is increasingly susceptible to the effects of global climate change. The 2022 report by the UN's Intergovernmental Panel on Climate Change warns that the PRC could be among the countries most severely impacted by rising global temperatures. Without adaptation and mitigation measures, the PRC's existing water scarcity issues will worsen, negatively affecting food security, increasing heat-related mortality rates, and placing terrestrial, coastal, and marine ecosystems at a higher risk of irreversible biodiversity loss and damage.[112] A separate analysis conducted by researchers from Peking University and the Chinese Academy of Sciences suggests that if climate-induced sea-level rise coincides with sudden-onset extreme storm surges, China's coastal regions, including mega-cities, such as Shanghai and Tianjin, could face an 11% loss of GDP by 2050.[113] Additionally, there are potential linkages between climate change and the PRC's national security. The *Third National Assessment Report on Climate Change*, published in 2015, suggests that diminishing river flows resulting from retreating glaciers in western China (the water source for Asia's major rivers) and ensuing disputes over water resources and transnational migration surges could lead to interstate conflicts on China's periphery.[114] To better prepare for and adapt to the effects of climate change, the Ministry of Ecology and Environment, along with 16 other departments, published the National Climate Change Adaptation Strategy 2035, which emphasises, among other things, climate change monitoring, early warning, and risk management.[115]

Regarding polar research, scientific evidence suggests possible connections between variations in the Arctic physical environment and weather anomalies in East Asia, including China.[116] Therefore, research on Arctic climate change and its impacts on China is a priority in China's Arctic scientific expeditions and a recurring theme in statements by CCP officials when discussing China's Arctic research and the proximity of the Chinese mainland to the Arctic region. For example, during the 2007–2008 International Polar Year (IPY), Chinese scientists launched the Arctic Change and its Tele-impacts on Mid-latitudes (ARCTIML) project to investigate Arctic environmental changes and assess their impact on the climate and weather in mid-latitudes, particularly in China.[117] More recently, some Chinese scientists have suggested connections between changes in the Arctic region and increased extreme cold surges during winter periods in China.[118] They have also observed that variations in Arctic Ocean surface temperature could contribute to dust generation in the Gobi Desert and create favourable meteorological conditions for its transport to north China, resulting in extreme events such as super sandstorms that impact agricultural production.[119] The Chinese Arctic White Paper acknowledges that changes in the Arctic "have a direct impact on China's climate system and ecological environment, and, in turn, on its economic interests in agriculture, forestry, fishery, marine industry, and other sectors."[120] Additionally, a 2017 study by researchers from the Georgia Institute of Technology implied a connection between China's winter haze and the melting of Arctic sea ice. These scientists suggested that receding Arctic sea ice led to increased evaporation and subsequent snowfall in the Arctic region, disrupting normal air pressure systems in Asia and altering the regular pathways of monsoon wind currents. This disruption hampers the natural ventilation process that helps clean the air in China, posing challenges for winter haze mitigation.[121] This may create some obstacles for the CCP leadership as it undercuts its efforts to combat air pollution and "make the sky blue again."[122] Consequently, Chinese scientific organisations will continue to be interested in the Arctic and collect data to understand and assess the impacts of Arctic environmental changes on the Chinese mainland.

Second, through scientific interests in the region, Chinese research institutions actively engage in international Arctic scientific cooperation and exchanges. These interactions form a cornerstone of Chinese Arctic activities. Chinese research institutions have participated in numerous international Arctic-related organisations, bodies, and committees, including the Ny-Ålesund Science Managers Committee, the International Arctic Science Committee, and the Sustaining Arctic Observing

Network, among others. China's international exposure was further enhanced through its active involvement in the fourth IPY, which included the ARCTIML project in the Arctic and the PANDA (Prydz Bay, Amery Ice Shelf, and Dome A Observations) in Antarctica.[123] Chinese officials also began to point out that the changes that were taking place in the Arctic were not merely regional but global in nature, influencing countries around the world, and for that reason more international cooperation was needed. For example, China's then Assistant Minister of Foreign Affairs, Hu Zhengyue, stated during the High North Study Tour in Svalbard that

> Arctic changes have global effects and can impact the socio-economic activities of states around the world, especially those in the Northern Hemisphere, requiring the full cooperation and joint resolve of the international community, and that is why the Chinese government pays high attention to Arctic affairs, actively conducting Arctic scientific research, and participating in international cooperation mechanisms.[124]

Ultimately, the CCP leveraged its science participation in Arctic research collaborations to gain access to the region and be recognised as a legitimate stakeholder in regional affairs.[125]

Furthermore, international Arctic scientific engagement has helped alleviate suspicions, to some extent, among certain Arctic states regarding China's presence in the region. Su and Mayer note that this was possible through the sharing of scientific resources, long-term interactions between Chinese and foreign scientists, and the reinforcement of Arctic knowledge-based institutions.[126] As a result, China could begin developing regional research networks such as the China-Nordic Arctic Research Centre in Shanghai. Established in 2013, the Centre serves as a platform for Arctic research cooperation, bringing together Chinese and Nordic research institutions. It focuses on Arctic climate change and its impacts, Arctic resources, shipping and economic cooperation, and Arctic policy and legal studies.[127] In addition, CNARC organises annual conferences, facilitates Chinese Nordic joint research projects, and supports visiting scholar exchanges. Its secretariat is located in Shanghai in the main PRIC building, and the PRIC director also serves as the director of CNARC. The impetus to establish such a research centre emerged after Hu Jintao visited Iceland in 2012 when the Chinese SOA and the Icelandic Ministry of Foreign Affairs signed an MoU on ocean and polar research.[128] Therefore, CNARC, particularly its annual conference, which alternate between China and selected Nordic countries, not only cultivates relationships between Arctic researchers but also connects government representatives and academics from China and Nordic countries.

In recent years, China has continued its participation in international Arctic scientific research, including the MOSAiC expedition – the largest Arctic expedition in history involving 600 experts from numerous countries – to study the interactions between the atmosphere, ocean, and sea ice.[129] Within the University of the Arctic (UArctic), a cooperative network of universities, research institutes, and other organisations focused on education and research in the North, there are 42 members from non-Arctic states as of 2023, with 15 of them being Chinese.[130] One of the Chinese members, the Harbin Institute of Technology, established the UArctic-HIT Training Centre to strengthen exchanges between Arctic and non-Arctic research organisations.[131] The importance of Arctic cooperation was also highlighted in China's 2018 Arctic White Paper, which listed it as one of the basic principles and effective means for the PRC's participation in Arctic affairs. Furthermore, Arctic scientific cooperation allows the PRC to engage with non-Arctic states interested in regional developments. In 2015, acknowledging the global relevance of the Arctic, the PRC, Japan and the Republic of Korea declared their intention to "launch a trilateral high-level dialogue on the Arctic to share Arctic policies, explore cooperative projects and seek ways to deepen cooperation over the Arctic."[132] The countries have already held four such dialogues in Seoul (2016), Tokyo (2017), Shanghai (2018), and Busan (2019).[133] These meetings reaffirm the importance of Arctic scientific cooperation, as the three countries support "the enhancement of the exchange of information on Arctic expeditions and encourage the sharing of scientific data and further development of collaborative surveys."[134]

Third, through its regional science activities, international Arctic scientific partnerships, relative geographical proximity to the Arctic, and emphasis on Arctic environmental changes and their impact on the PRC's ecological environment, the CCP has been able to establish China's Arctic credentials and develop a (near) Arctic identity. Marc Lanteigne from UiT The Arctic University of Norway notes that while Asia-Arctic nations like China have some Arctic-related experiences, they lack an extensive historical legacy of far-north exploration around which they could construct their Arctic identity (unlike some European observers to the AC). Therefore, they have had to devise alternative means of identity-building. In this case, they turned to science collaborations, developing regional science interests in partnerships with Arctic states while also highlighting the potential of regional economic development.[135]

To further justify its interests in the Arctic, the PRC reinforced its near-Arctic identity by incorporating the notion of being a near-Arctic state into its official documents and statements.[136] Deng explains that a near-Arctic identity entails close proximity to the Arctic region, convenient transportation access, a direct link between the region and the state's climatic, environmental, and ecological conditions, and a sustained commitment to engaging in Arctic affairs.[137] This identity fosters a perception that the state has a natural responsibility to participate in Arctic affairs. However, despite PRC analysts claiming that a near-Arctic identity does not challenge the dominant position of Arctic states in regional affairs,[138] China faced opposition from some Arctic states, particularly the US. In 2019, then US Secretary of State, Mike Pompeo, dismissed China's claims of being a near-Arctic state, stating that "there are Arctic states and non-Arctic states. No third category exists. China claiming otherwise entitles them to exactly nothing."[139]

Meanwhile, as the PRC expands its regional involvement beyond science and scientific collaborations, its Arctic policy maintains China's significance as an important Arctic stakeholder. Some Chinese analysts view this as another aspect of Arctic identity formation. While near-Arctic status pertains to geographical proximity, an Arctic stakeholder identity specifically refers to the depth, variety, and different dimensions of a state's Arctic interests.[140] According to Deng, what drives the construction of this identity is opposition by non-Arctic states to the perceived dominance of Arctic states over all aspects of regional affairs based on nothing more that the particularity of their geographical location in the Arctic region.[141]

In conclusion, the development of Chinese Arctic scientific infrastructure, research organisations, and hardware signifies the commitment of the Party-state leadership to enhance its understanding of the Arctic and safeguard its ecological security. Arctic science also enables the Chinese scientific community to participate in collaborative projects, expanding China's global knowledge networks and, to some extent, mitigating concerns about Chinese scientific presence in the region. However, under Xi Jinping's leadership, China has embraced a comprehensive security perspective to guide its development, introducing new dimensions that should be considered in its Arctic engagement beyond scientific observations for ecological security. These dimensions will be explored in the next chapter.

Notes

1 "中华人民共和国国民经济和社会发展第十四个五年规划和2035年远景目标纲要" [Outline of the PRC's 14th Five-Year Plan for National Economic and Social Development and Long-Term Objectives for 2035], *Xinhua*, March 3, 2021, www.gov.cn/xinwen/2021-03/13/content_5592681.htm.

2 Liu Shiping, "雪龙探极：提升我国极地工作5大能力" [Xuelong Exploring the Polar Regions: Improving 5 Major Capabilities of My Country's Polar Work], *Xinhua*, October 11, 2018, www.gov.cn/xinwen/2018-10/11/content_5329708.htm.

3 "Keynote Speech by Vice Foreign Minister Zhang Ming at the China Country Session of the Third Arctic Circle Assembly," *Ministry of Foreign Affairs of the PRC*, October 17, 2015, www.fmprc.gov.cn/mfa_eng/wjdt_665385/zyjh_665391/201510/t20151017_678393.html.

4 Qu Tanzhou et al., 北极问题研究 [*Research on Arctic Issues*] (Beijing: 海洋出版社 [Ocean Press], 2011), 131–188.

5 Zuo Pengfei, 极地战略问题研究 [*A Study on Polar Strategy*] (Beijing: 时事出版社 [Current Affairs Press], 2018), 7–9.

6 Zhang Baoshu, "北极无人冰站，中国造！" [Arctic Unmanned Ice Station, Made in China!], 人民网 [*People's Daily Online*], October 17, 2018, https://web.archive.org/web/20221007133738/http://scitech.people.com.cn/n1/2018/1017/c1007-30345336.html.

7 "中心职责" [Institute's Responsibilities], 中国极地研究中心 [*The Polar Research Institute of China*], n.d., www.pric.org.cn/index.php?c=category&id=4.

8 See "科学研究" [Polar Research], 中国极地研究中心 [*The Polar Research Institute of China*], n.d., www.pric.org.cn/index.php?c=category&id=25.

9 "极地政策研究室" [Division of Polar Policy Studies], 中国极地研究中心 [*The Polar Research Institute of China*], n.d., www.pric.org.cn/index.php?c=category&id=43.

10 "重点实验室" [Key Laboratory of Polar Science], 中国极地研究中心 [*The Polar Research Institute of China*], n.d., www.pric.org.cn/index.php?c=category&id=25.

11 "期刊简介" [About the Journal], 中国极地研究中心 [*The Polar Research Institute of China*], December 28, 2021, www.pric.org.cn/index.php?c=show&id=522.

12 See also Anne-Marie Brady, *China as a Polar Great Power* (Cambridge: Cambridge University Press, 2017), 130.

13 "第五届中国 – 北欧北极合作研讨会在大连海事大学开幕" [The Fifth China-Nordic Arctic Cooperation Symposium Was Held at Dalian Maritime University], 大连海事大学 [*Dalian Maritime University*], May 31, 2017, www.dlmu.edu.cn/info/1089/2675.htm.

14 "极地大科学与工程研究中心助力北极航行" [The Polar Science and Engineering Research Centre Assists in Arctic Shipping], 哈尔滨工程大学 [*Harbin Engineering University*], April 19, 2017, http://sec.hrbeu.edu.cn/2017/0419/c441a115866/page.htm.

15 "工信部：哈工程北极蓝色经济研究中心揭牌" [Ministry of Industry and Information Technology: HEU's Arctic Blue Economy Research Centre Unveiled], 中华人民共和国工业和信息化部 [*Ministry of Industry and Information Technology of the PRC*], November 23, 2018, http://news.hrbeu.edu.cn/info/1017/2969.htm.

16 "研究院简介" [Academy Introduction], 哈尔滨工业大学 [*Harbin Institute of Technology*], n.d., http://polar.hit.edu.cn/11640/list.htm.

17 "About Us," *Arctic Studies Centre*, n.d., http://en.asclcu.cn/about/index_1.asp.

18 "团队概况" [Team Profile], 中国海洋大学海洋发展研究院 [*Institute of Marine Development of Ocean University of China*], n.d., http://hyfzyjy.ouc.edu.cn/jdyjzx/list.html.

19 "极地与深海发展战略研究中心" [Centre for Polar and Deep Ocean Development], 上海交通大学凯原法学院 [*Koguan School of Law, Shanghai Jiaotong University*], n.d., https://law.sjtu.edu.cn/hyflyzfyjzx/index.html.

20 "学院简介" [School Profile], 中山大学测绘科学与技术学院 [*Sun Yat-sen University, School of Geospatial Engineering and Science*], n.d., https://sges.sysu.edu.cn/about.

21 "同济大学极地与海洋国际问题研究中心简介" [Introduction to the Centre for Polar and Oceanic Studies at Tongji University], 同济大学 [*Tongji University*], n.d., https://cpos.tongji.edu.cn/17269/list.htm.

22 "学院简介" [School Profile], 武汉大学中国边界与海洋研究院 [*China Institute of Boundary and Ocean Studies of Wuhan University*], n.d., www.cibos.whu.edu.cn/index.php?id=about.

23 Universities included: Harbin Engineering University, Jilin University, Dalian University of Technology, Dalian Maritime University, Tsinghua University, Peking University, Beijing Normal University, Taiyuan University of Technology, China Ocean University, Shandong University, Lanzhou University, Nanjing University, Hohai University, Nanjing University of Information Science and Technology, Tongji University, Fudan University, Shanghai Maritime University, East China Normal University, Shanghai Jiaotong University, University of Science and Technology of China, Wuhan University, Wuhan University of Technology, Huazhong University of Science and Technology, Xiamen University, and Zhongshan University; see "我校成为"中国高校极地联合研究中心"共建单位" [Our University Has Become One of the Founding Units of the Joint Centre for Polar Research of Chinese Universities], 太原理工大学 [*Taiyuan University of Technology*], April 23, 2018, http://dlxy.tyut.edu.cn/info/1064/1564.htm.

24 Brady, *China as a Polar Great Power*, 130.

25 "海洋和极地研究中心" [Centre for Marine and Polar Studies], 上海国际问题研究院 [*Shanghai Institutes for International Studies*], n.d., www.siis.org.cn/research/12.jspx.

26 "我国极地气象考察现状" [Current Situation of Polar Meteorological Observations in China], 中国气象科学研究院 [*Chinese Academy of Meteorological Sciences*], n.d., www.camscma.cn/article/56.html.

27 For example, see: "中国科学院大气物理研究所" [The Institute of Atmospheric Physics of the Chinese Academy of Sciences Launched the 2018 Arctic Atmospheric Science Experiment], 中国科学院大气物理研究所 [*The Institute of Atmospheric Physics, Chinese Academy of Sciences*], June 6, 2018, https://web.archive.org/web/20181219140806/www.iap.ac.cn/xwzx/tpxw/201806/t20180604_5020968.html.

28 Gao Huili and Yang Jian, "中国地质调查局极地地学研究中心成立" [The Establishment of the Polar Geology Research Centre at the China Geological Survey], 中国自

然资源报 [*China Natural Resources News*], November 2, 2020, www.cgs.gov.cn/xwl/kxjs/202011/t20201102_657953.html.

29   For example, see: Zhang Xudong, "向阳红01船起航执行中国第10次北极考察" [The Ship Xiang Yang Hong 1 Set Sail for China's 10th Arctic Expedition], *Xinhua*, August 10, 2019, www.xinhuanet.com/politics/2019-08/10/c_1124860463.htm.

30   Iselin Stensdal, "Coming of Age? Asian Arctic Research, 2004–2013," *Polar Record* (2015): 3.

31   "北极黄河站" [Arctic Yellow River Station], 中国极地研究中心 [*The Polar Research Institute of China*], n.d., www.pric.org.cn/index.php?c=category&id=98.

32   "UK Arctic Research Station," *Arctic Office of the Natural Environment Research Council*, n.d., www.arctic.ac.uk/uk-arctic-research-station/.

33   "Ny-Ålesund Science Plan," *National Centre for Antarctic and Ocean Research*, n.d., www.ncaor.gov.in/arctics/display/123-ny-alesund-science-plan.

34   Chinese Arctic and Antarctic Administration, *2015 National Annual Report on Polar Program* (Beijing, China, December 2015), 18–21.

35   "Cornerstone Laid for Chinese-Icelandic Aurora Observatory (CIAO) in Kárhóll," *Arctic Portal*, October 12, 2016, https://arcticportal.org/ap-library/news/1794-cornerstone-laid-for-chinese-icelandic-aurora-observatory-ciao-in-karhol.

36   "中冰北极科学考察站" [China-Iceland Arctic Science Observatory], 中国极地研究中心 [*The Polar Research Institute of China*], n.d., www.pric.org.cn/index.php?c=category&id=99.

37   Zhao Ning, "海洋局：中冰联合极光观测台建设取得重要进展" [State Oceanic Administration: Significant Progress Was Made in the Construction of the Chinese-Icelandic Aurora Observatory], 国家海洋局 [*The State Oceanic Administration*], October 26, 2016, www.gov.cn/xinwen/2016-10/26/content_5124514.htm.

38   "China-Iceland Arctic Science Observatory Inaugurated in Northern Iceland," *Xinhua*, October 19, 2018, www.xinhuanet.com/english/2018-10/19/c_137542493_2.htm.

39   Vala Hafstad, "Chinese Research Facility on Icelandic Farmland," *Icelandic Review*, July 13, 2016, https://web.archive.org/web/20170609235310/http://icelandreview.com/news/2016/07/13/chinese-research-facility-icelandic-farmland.

40   Lu Junyuan and Zhang Xia, 中国北极权益与政策研究 [*China's Arctic Interests and Policy*] (Beijing: 时事出版社 [Current Affairs Press], 2016), 236.

41   "海洋局副局长出席《中华人民共和国国家海洋局和格陵兰教育、文化、研究和宗教部科学合作谅解备忘录》签字仪式和拜会丹麦外交部北极大使" [The Deputy Director of the State Oceanic Administration Attended the Signing Ceremony of the Memorandum of Understanding on Scientific Cooperation Between the State Oceanic Administration of the People's Republic of China and the Greenlandic Ministry of Education, Culture, Research and Church, and Also Visited the Arctic Ambassador of the Danish Foreign Ministry], 国家海洋局 [*State Oceanic Administration*], May 16, 2016, www.gov.cn/xinwen/2016-05/16/content_5073689.htm.

42   Jichang Lulu, "China Wants Greenland Station ASAP; One Candidate Site Near Planned China Nonferrous Investment," *Worldpress (blog)*, October 14, 2017, https://jichanglulu.wordpress.com/2017/10/14/china-wants-greenland-station-asap-one-candidate-site-near-planned-china-nonferrous-investment/.

43   "雪龙号" [R/V Xuelong], 国家海洋局极地考察办公室 [*Chinese Arctic and Antarctic Administration*], October 14, 2022, http://chinare.mnr.gov.cn/catalog/detail?id=4fd5c71775ef4cd78ef396049e7efb95&from=zzjgkcc&currentIndex=1.

44   Chinese Arctic and Antarctic Administration (CAA), *2007 National Annual Report on Polar Program of China* (Beijing, China, March 2008), 61–63, and Chinese Arctic and Antarctic Administration (CAA), *2013 National Annual Report on Polar Program of China* (Beijing, China, December 2013), 75.

45   Luo Sha, "我国新建极地科学考察破冰船力争2013年投入使用" [China Will Construct a New Polar Research Icebreaker, Plans to Launch It in 2013], *Xinhua*, June 21, 2011, www.gov.cn/jrzg/2011-06/21/content_1889618.htm.

46 Cui Shuang, "世界首艘双向破冰极地科考船！"雪龙2"号技能满格" [The World's First Two-Directional Ice-Breaking Polar Research Vessel! Xuelong 2 Packed with Capability], 科技日报 [*Science and Technology Daily*], June 29, 2021, https://m.gmw.cn/baijia/2021-06/29/1302381762.html.

47 "我们的企业" [Our Business], 中国船舶集团有限公司 [*China State Shipbuilding Corporation Limited*], n.d., www.cssc.net.cn/n4/n12/index.html.

48 "雪龙2号" [Xuelong 2], 国家海洋局极地考察办公室 [*Chinese Arctic and Antarctic Administration*], September 26, 2022, http://chinare.mnr.gov.cn/catalog/detail?id=22cd30d7444f4f4399341ad5586768ed&from=zzjgkcc&currentIndex=1.

49 Deng Qi, "雪龙2号下水 可双向破厚冰实现非夏季极地考察" [The Launch of Xue-long 2. It Can Break Thick Ice Bi-Directionally to Conduct Non-summer Polar Ex-plorations], 新京报 [*The Beijing News*], September 11, 2018, https://baijiahao.baidu.com/s?id=1611265897843864957&wfr=spider&for=pc.

50 Cui, "世界首艘双向破冰极地科考船！"雪龙2"号技能满格" [The World's First Two-Directional Ice-Breaking Polar Research Vessel! Xuelong 2 Packed with Capability].

51 Bryan J.R. Millard and P. Whitney Lackenbauer, *Trojan Dragons? Normalizing Chi-na's Presence in the Arctic* (Calgary: Canadian Global Affairs Institute, 2021), 14.

52 Lu and Zhang, 中国北极权益与政策研究 [*China's Arctic Interests and Policy*], 240.

53 Bai Guolong and Zhang Jiansong, "国家海洋局：我国北极科考频次增至每年一次" [State Oceanic Administration: The Frequency of China's Arctic Scientific Ex-amination Has Increased to One Per Year], *Xinhua*, October 10, 2017, www.xinhuanet.com/politics/2017-10/10/c_1121781250.htm.

54 Yu Qiongyuan, "中国首次成功试航北极西北航道" [China Successfully for the First Time Navigated the Arctic Northwest Passage], *Xinhua*, September 7, 2017, www.xinhuanet.com/politics/2017-09/07/c_1121625642.htm.

55 For example, see: Xu Ren, ed., 中国第八次北极科学考察报告 [*The Report of 2017 Chinese National Arctic Research Expedition*] (Beijing: 海洋出版社 [Ocean Press], 2019).

56 Qu Jing, "我国首次在北极布放长期自动气象观测系统" [China for the First Time Deployed a Long-Term Automatic Meteorological Observation System in the Arctic], *Xinhua*, September 6, 2012, www.gov.cn/jrzg/2012-09/06/content_2218615.htm.

57 Zheng Tianhao, "我国在北极P冰川成功架设首台自动气象站" [China Suc-cessfully Set Up the First Automatic Weather Station in P Glacier in the Arc-tic], 央视新闻移动网 [*News CCTV*], May 16, 2019, www.bjnews.com.cn/news/2019/05/16/579765.html.

58 "重新连接！这台气象站漂泊600多天后续写传奇！" [Reconnected! This Weather Station Has Been Drifting for More than 600 Days Days], 中国气象微信公众号 [*China Meteorology Official WeChat Account*], December 2, 2021, www.stdaily.com/cehua/Dec2nd/2021-12/02/content_1235788.shtml.

59 Zheng Tianhao and Li Jie, "我国冰基漂流自动气象站首次在北极布放成功" [Chi-na's Ice Drifting Automatic Weather Station Was Successfully Deployed in the Arctic], 央视网 [*CCTV.com*], September 5, 2018, http://news.china.com.cn/2018-09/05/content_62344176.htm.

60 "中国气象局 科学技术部 中国科学院 关于印发《中国气象科技发展规划（2021–2035年）》的通知" Notification by the China Meteorological Administration, the Min-istry of Science and Technology and the Chinese Academy of Science on the publication of the 2021–2035 China Meteorological Science and Technology Development Plan], 广东省气象局 [*Guangdong Meteorological Administration*], February 28, 2022, http://gd.cma.gov.cn/zfxxgk/zwgk/ghjh/202203/t20220303_4555674.html.

61 "About EISCAT," *EISCAT Scientific Association*, November 21, 2022, www.eiscat.se/about/.

62 "企业简介" [Company Introduction], 中国电子科技集团公司第22研究所 [*China Electronics Technology Group Corporation No. 22 Research Institute*], n.d., https://web.archive.org/web/20220107224356/www.crirp.ac.cn/22/338987/338975/index.html.

63 Nina Berglund, "Norway Won't Let China Build a Radar," *NewsinEnglish.no*, September 12, 2014, www.newsinenglish.no/2014/09/12/norway-wont-let-china-build-radar/.

64 Qu Jing, "中国北极科考队布放首个极地大型海洋观测浮标" [China's Arctic Research Team Has Launched Its First Large Polar Ocean Observation Buoy], *Xinhua*, August 5, 2012, www.gov.cn/jrzg/2012-08/05/content_2198526.htm.

65 "中国第七次北极科考队投放锚定浮标" [China's Seventh Arctic Scientific Expedition Deployed a Moored Buoy], *Xinhua*, July 20, 2016, www.gov.cn/xinwen/2016-07/20/content_5093071.htm.

66 Chen Yu, "北极海域有了中国的拖曳式冰浮标" [China's Towed Ice Buoy in Arctic Waters], 科技日报 [*Science and Technology Daily*], August 19, 2014, https://web.archive.org/web/20220920133914/http://scitech.people.com.cn/n/2014/0819/c1057-25492226.html.

67 Chen Chao et al., "极地海冰浮标观测技术" [Polar Sea Ice Buoy Observation Technology], 科学 [*Science*] no. 3 (2021), www.fx361.com/page/2021/1201/9171791.shtml.

68 "海上观测平台" [Ocean Observation], 中国海洋大学 [*Ocean University of China*], n.d., https://pol.ouc.edu.cn/hsgcptsjgxpt/list.htm.

69 "我校博士研究生杨波参加中国第12次北极科学考察" [Yang Bo, a Doctoral Student from Our University, Participated in China's 12th Arctic Scientific Expedition], 太原理工大学 [*Taiyuan University of Technology*], July 13, 2021, https://ices.tyut.edu.cn/info/1046/3202.htm.

70 "中国第九次北极科考圆满完成 首次布放无人冰站" [China's Ninth Arctic Scientific Expedition Successfully Deployed the First Unmanned Ice Station], 中国青年报 [*China Youth Daily*], October 14, 2018, https://web.archive.org/web/20181014060413/www.xinhuanet.com/science/2018-10/14/c_137531200.htm.

71 Chen et al., "极地海冰浮标观测技术" [Polar Sea Ice Buoy Observation Technology].

72 Shen Cheng, "特稿：中国北极冰站观测迈入'无人时代'" [Special Report: Chinese Arctic Ice Station Observations Usher in the Unmanned Era], *Xinhua*, August 22, 2018, www.xinhuanet.com/politics/2018-08/22/c_129938144.htm.

73 Zeng Junbao, Li Shuo and Liu Ya, "Application of Unmanned Underwater Vehicles in Polar Research," *Advances in Polar Science* 32, no. 3 (2021): 179–180.

74 "北极ARV" [Arctic ARV], 沈阳自动化研究所 [*Shenyang Institute of Automation*], December 13, 2015, www.cas.cn/kx/kpwz/201512/t20151213_4493176.shtml.

75 Wei Zexun, ed., 中国第九次北极科学考察报告 [*The Report of 2018 Chinese Arctic Research Expedition*] (Beijing: 海洋出版社 [Ocean Press], 2019), 26.

76 Wu Yuehui, "我国研发的自主水下机器人首次完成北极科考" [An Autonomous Underwater Robot Developed by China Completed Its First Arctic Scientific Expedition], 人民日报 [*People's Daily*], October 25, 2021, www.chinanews.com.cn/gn/2021/10-25/9594244.shtml.

77 Cheng Xiao et al., "极地环境探索关键技术" [Critical Technologies for Detection of the Polar Environment], 中国科学院院刊 [*Bulletin of Chinese Academy of Sciences*], 37, no. 7 (2022): 925.

78 Fang Zhengfei, "专家详解我国卫星极地观测应用发展" [Experts Explain the Development of Satellite Polar Observation Applications], 中国自然资源报 [*China Natural Resources News*], June 7, 2022, www.mnr.gov.cn/dt/hy/202206/t20220607_2738578.html.

79 Chen Liangfu et al., "An Introduction to the Chinese High-Resolution Earth Observation System: Gaofen 1–7 Civilian Satellites," *Journal of Remote Sensing* (2022): 3, https://doi.org/10.34133/2022/9769536.

80 Cheng et al., "极地环境探索关键技术" [Critical Technologies for Detection of the Polar Environment], 926.

81 William Graham, "China Launches Ziyuan-1 02E Satellite via Chang Zheng 4C," *NASA Spaceflight*, December 26, 2021, www.nasaspaceflight.com/2021/12/china-ziyuan-1-02e/#:~:text=China%20has%20deployed%20a%20new,11%3A11%20Beijing%20Time.

82 Cheng et al., "极地环境探索关键技术" [Critical Technologies for Detection of the Polar Environment], 926.

83 "FY-3 Series," *National Satellite Meteorological Centre*, n.d., www.nsmc.org.cn/nsmc/en/satellite/FY3.html.

84 Cheng et al., "极地环境探索关键技术" [Critical Technologies for Detection of the Polar Environment], 926.

85 Ibid.

86 Zhang Guohang, Mao Lingye and Fu Yifei, "20年, 中国海洋卫星服务遍及全球" [20 Years, China's Oceanic Satellite Service Spans the Globe], 科技日报 [*Science and Technology Daily*], May 15, 2022, www.stdaily.com/index/kejixinwen/202205/6ee61c360e3e4ff28229092f5369ea92.shtml.

87 Zhang Ying et al., "Accuracy Evaluation on Geolocation of the Chinese First Polar Microsatellite (Ice Pathfinder) Imagery," *Remote Sensing* 13, no. 4278 (2021): 2.

88 Dan Ni, "冰路飞天, 守望"三极" – 专访冰路卫星项目发起人, 中山大学测绘科学与技术学院院长程晓" [Ice Pathfinder Flying in the Sky, Observing the Three Poles – An Interview with Cheng Xiao, the Dean of School of Geospatial Engineering and Science at Sun Yat-Sen University, the Initiator of Ice Pathfinder Satellite Project], 中国测绘 [*China Surveying and Mapping*], no. 10 (2019): 19–22, https://m.thepaper.cn/baijiahao_5258444?sdkver=e06426d6&clientprefetch=1.

89 Yang Shuxin, "北极航道监测科学试验卫星计划2022年发射" [The Arctic Shipping Route Monitoring Scientific Experimental Satellite Is Scheduled to Be Launched in 2022], Xinhua, December 4, 2020, www.xinhuanet.com/politics/2020-12/04/c_1126823230.htm.

90 Chen Yu, "冰路卫星下月飞天 专家详解三极遥感星座观测系统" [The Ice Pathfinder Satellite Is Set to Fly Next Month, Experts Detailing the Tripolar Remote Sensing Constellation Observation System], 科技日报 [*Science and Technology Daily*], August 19, 2019, http://ccnews.people.com.cn/n1/2019/0819/c141677-31302636.html.

91 Fang, "专家详解我国卫星极地观测应用发展" [Experts Explain the Development of Satellite Polar Observation Applications].

92 Since then, it seems that the institute has been merged with two other branches of CAS to create the CAS Aerospace Information Research Institute (中国科学院空天信息创新研究院), see "欢迎访问中国科学院空天信息创新研究院" [Welcome to the CAS Aerospace Information Research Institute], 中国科学院空天信息创新研究院 [*CAS Aerospace Information Research Institute*], n.d., www.aircas.cas.cn/index_73758.html.

93 "China's First Overseas Land Satellite Receiving Station Put Into Operation," *Chinese Academy of Sciences*, December 16, 2016, https://english.cas.cn/newsroom/archive/news_archive/nu2016/201612/t20161215_172471.shtml.

94 "遥感地球所与芬兰气象研究所签订中芬北极空间观测联合研究合作协议" [The Institute of Remote Sensing and the Finnish Meteorological Institute Signed a Sino-Finnish Arctic Space Observation Joint Research Cooperation Agreement], 中国科学院 [*Chinese Academy of Sciences*], April 12, 2018, www.cas.cn/yx/201804/t20180417_4642339.shtml. The agreement for the joint operation of the research centre was not renewed and terminated in 2021. Reports indicate that this was due to changes in the geopolitical situation as well as disinterest from both parties. See Michael Lipin, "China Begins to Revive Arctic Scientific Ground Projects After Setbacks," *VOA*, December 5, 2022, www.voanews.com/a/china-begins-to-revive-arctic-scientific-ground-projects-after-setbacks-/6860756.html.

95 "中俄地面站合作开展北极航道卫星遥感监测" [Chinese and Russian Ground Stations Cooperation in Satellite Remote Sensing Monitoring of Arctic Shipping Routes], 中国科学院空天信息创新研究院 [*Aerospace Information Research Institute, Chinese Academy of Sciences*], July 26, 2019, www.aircas.cas.cn/dtxw/kydt/201907/t20190726_5351552.html.

96 Cheng et al., "极地环境探索关键技术" [Critical Technologies for Detection of the Polar Environment], 927.

97 Ibid.
98 Liu Yang, "中国遥感无人机格陵兰岛首飞成功" [China's Remote Sensing Drone Has Made Its Maiden Flight in Greenland], 环球网 [*Global Times*], June 7, 2017, https://uav.huanqiu.com/article/9CaKrnK3fPI.
99 国家海洋局 [State Oceanic Administration], 2013 年度中国极地考察报告 [*2013 Annual Report on Chinese Polar Research*] (Beijing, China, December 2013), 59.
100 国家海洋局 [State Oceanic Administration], 2015 年度中国极地考察报告 [*2015 Annual Report on Chinese Polar Research*] (Beijing, China, December 2015), 56–58.
101 Wu Nan, "杨惠根: 极地考察期待更多工程技术研究" [Yang Huigen: Polar Expeditions to Include More Engineering Research], 中国社会科学网 [*Chinese Social Science Net*], April 16, 2015, https://web.archive.org/web/20210706114354/www.cssn.cn/gd/gd_rwhd/xslt/201504/t20150416_1589295.shtml.
102 Sun Kai, "Beyond the Dragon and the Panda: Understanding China's Engagement in the Arctic," *Asia Policy* 18 (2014): 47.
103 The author's own search through the database of the National Social Science Fund at: http://fz.people.com.cn/skygb/sk/index.php/Index/seach.
104 Zhao Ning, "国家海洋局副局长林山青解读《北极考察活动行政许可管理规定》" [Lin Shanqing, the Deputy Director of the State Oceanic Administration Interprets the Administrative Licensing Regulations for Arctic Research Activities], 中国海洋报 [*China Ocean News*], September 20, 2017, https://web.archive.org/web/20171120205440/www.soa.gov.cn/zwgk/zcjd/201709/t20170920_57973.html.
105 An Haiyan, "我们为什么重视极地考察" [Why Do We Attach Importance to Polar Exploration], 中国海洋报 [*China Ocean News*], September 12, 2017, https://web.archive.org/web/20190101051648/www.oceanol.com/zhuanti/201709/12/c68270.html.
106 "海洋局关于印发《北极考察活动行政许可管理规定》的通知" [Notification by the State Oceanic Administration on the Publication of Administrative Licensing Regulations for Arctic Research Activities], 国家海洋局 [*State Oceanic Administration*], July 7, 2017, www.gov.cn/xinwen/2017-09/07/content_5223192.htm.
107 Wu, "杨惠根: 极地考察期待更多工程技术研究" [Yang Huigen: Polar Expeditions to Include More Engineering Research].
108 Liu Shiping, "中国极地科学技术委员会正式成立" [Chinese Polar Science and Technology Commission Was Established], *Xinhua*, December 28, 2017, https://web.archive.org/web/20200811081939/www.xinhuanet.com/2017-12/28/c_1122181477.htm.
109 Chen Yu, "十三五我国将在北极新建考察站" [13th Five-Year Plan – China Will Construct a New Arctic Research Station], 科技日报 [*Science and Technology Daily*], December 29, 2017, http://news.sciencenet.cn/htmlnews/2017/12/398582.shtm.
110 Sun Qiuci, "我所主导的北极JASMInE科学计划圆满完成" [My Institute's Arctic JASMInE Science Program Successfully Completed], 自然资源部第二海洋研究所 [*Second Institute of Oceanography, MNR*], October 15, 2021, https://web.archive.org/web/20220405112717/www.sio.org.cn/redir.php?catalog_id=84&object_id=330453.
111 "总体国家安全观的16种安全" [16 Kinds of Security of the Comprehensive National Security Outlook], 国安宣工作室 [*National Security Propaganda Office*], April 14, 2021, www.stdaily.com/cehua/20210414/2021-04/14/content_1114342.shtml.
112 Yuan Ye, "IPCC Warns China Will Be Hit Hard by Climate Change," *The Sixth Tone*, March 3, 2022, www.sixthtone.com/news/1009809/ipcc-warns-china-will-be-hit-hard-by-climate-change#:~:text=Water%20and%20food%20insecurity%20are,risks%20in%20a%20warmer%20future. For the IPCC report see Rajib Shaw et al., "Asia," in *Climate Change 2022: Impacts, Adaptation and Vulnerability. Contribution of Working Group II to the Sixth Assessment Report of the Intergovernmental Panel on Climate Change*, ed. H.-O. Pörtner, D.C. Roberts, M. Tignor, E.S. Poloczanska, K. Mintenbeck, A. Alegría, M. Craig, S. Langsdorf, S. Löschke, V. Möller, A. Okem, B. Rama (Cambridge and New York: Cambridge University Press, 2022): 1457–1579, www.ipcc.ch/report/ar6/wg2/downloads/report/IPCC_AR6_WGII_Chapter10.pdf.
113 Cui Qi, Xie Wei and Liu Yu, "Effects of Sea Level Rise on Economic Development and Regional Disparity in China," *Journal of Cleaner Production* 176 (2018): 1245–1253.

114   Chris Buckley, "Chinese Report on Climate Change Depicts Sombre Scenarios," *The New York Times*, November 29, 2015, www.nytimes.com/2015/11/30/world/asia/chinese-report-on-climate-change-depicts-somber-scenarios.html?mcubz=3.

115   "生态环境部等17部门联合印发《国家适应气候变化战略2035》" [Ministry of Ecology and Environment, 17 Departments Jointly Issue the National Climate Change Adaptation Strategy 2035], 中华人民共和国生态环境部 [*Ministry of Ecology and Environment of the PRC*], June 23, 2022, https://web.archive.org/web/20230202213506/www.mee.gov.cn/ywdt/xwfb/202206/t20220613_985408.shtml.

116   Jin-Soo Kim et al., "Arctic Warming-Induced Cold Damage to East Asian Terrestrial Ecosystems," *Communications Earth & Environment* 3, no. 16 (2022): 1–8.

117   Yang Huigen, "Development of China's Polar Linkages," *Canadian Naval Review* 8, no. 3 (2012): 31.

118   Zheng Fei et al., "The 2020/21 Extremely Cold Winter in China Influenced by the Synergistic Effect of La Niña and Warm Arctic," *Advances in Atmospheric Sciences* 39, (2022): 546–552.

119   Zhang Nannan, "Ocean Variability Contributes to Sandstorms in Northern China," *Phys.org*, August 1, 2022, https://phys.org/news/2022-08-ocean-variability-contributes-sandstorms-northern.html.

120   "Full Text: China's Arctic Policy," *Xinhua*, January 26, 2018, www.xinhuanet.com/english/2018-01/26/c_136926498.htm.

121   Zou Yufei et al., "Arctic Sea Ice, Eurasia Snow, and Extreme Winter Haze in China," *Science Advances* 3, no. 3 (2017), http://advances.sciencemag.org/content/3/3/e1602751.full.

122   Patrick Reilly, "How Melting Arctic Sea Ice Is Keeping Smog Over China," *The Christian Science Monitor*, March 16, 2017, www.csmonitor.com/Environment/2017/0316/How-melting-Arctic-sea-ice-is-keeping-smog-over-China.

123   Tang Yao, "Development of the International Polar Years and Their Benefit for China," *Advances in Polar Sciences* 33, no. 2 (2022): 192–198.

124   Hu Zhengyue as cited in Ning Xiaoxiao, "地球未来的缩影-外交部部长助理胡正跃谈北极研究之旅" [A Microcosm of the World's Future – Assistant Minister of Foreign Affairs Hu Zhengyue Talks about High North Study Tour], 世界博览 [*World Vision*] 349, no. 19 (2009): 59.

125   Rasmus Gjedssø Bertelsen, Li Xing and Mette Højris Gregersen, "Chinese Arctic Science Diplomacy: An Instrument for Achieving the Chinese Dream?" in *Global Challenges in the Arctic Region: Sovereignty, Environment and Geopolitical Balance*, eds. Elena Conde and Sara Iglesias Sánchez (Oxford: Taylor & Francis, 2016), 442.

126   Su Ping and Maximilian Mayer, "Science Diplomacy and Trust Building: 'Science China' in the Arctic," *Global Policy* 9, no. 3 (2018): 24–26.

127   "Organization," *China-Nordic Arctic Research Centre*, n.d., www.cnarc.info/organization.

128   Chou Jiangtao, "中国 – 北欧北极研究中心成立" [China-Nordic Arctic Research Centre Established], 国家海洋局 [*The State Oceanic Administration*], December 25, 2013, www.mnr.gov.cn/dt/hy/201312/t20131225_2331697.html.

129   Wang Yan, "Arctic Exploration: Drifting with the Ice," *China Dialogue Ocean*, March 2, 2020, https://chinadialogueocean.net/en/climate/13226-arctic-exploration-drifting-with-the-ice/.

130   "Non-Arctic," *UArctic*, n.d., www.uarctic.org/members/member-profiles/non-arctic/.

131   "北极大学联盟-哈工大培训中心简介" [Introduction to UArctic – HIT Training Centre], 哈尔滨工业大学极地研究院 [*Polar Academy Harbin Institute of Technology*], n.d., http://polar.hit.edu.cn/bjdxlmwhgdpxzx/list.htm.

132   "Joint Declaration for Peace and Cooperation in Northeast Asia," *Ministry of Foreign Affairs of Japan*, November 1, 2015, www.mofa.go.jp/a_o/rp/page1e_000058.html.

133   "The Fourth Trilateral High-Level Dialogue on the Arctic, Busan, June 25–26, 2019," *Ministry of Foreign Affairs of the Republic of Korea*, June 27, 2019, www.mofa.go.kr/eng/brd/m_5676/view.do?seq=320574.

134 "Joint Statement the Third Trilateral High-Level Dialogue on the Arctic," *Ministry of Foreign Affairs of the PRC*, June 8, 2018, www.fmprc.gov.cn/mfa_eng/wjdt_665385/2649_665393/201806/t20180608_679524.html.
135 Marc Lanteigne, "Walking the Walk: Science Diplomacy and Identity-Building in Asia-Arctic Relations," *Jindal Global Law Review* 8, no. 1 (2017): 89–90.
136 The idea of China being a near-Arctic state was originally introduced by Zhang Xia from the PRIC and later adopted by CCP officials. For the original source, see Lu Junyuan, 北极地缘政策与中国应对 [*Geopolitics in the Arctic and China's Response*] (Beijing: 时事出版社 [Current Affairs Press], 2010), 339.
137 Deng Beixi, 北极安全研究 [*Arctic Security Studies*] (Beijing: 海洋出版社 [Ocean Press], 2020), 197.
138 Ibid.
139 "US Stuns Audience by Tongue-Lashing China, Russia on Eve of Arctic Council Ministerial," *The Barents Observer*, May 6, 2019, https://thebarentsobserver.com/en/arctic/2019/05/us-stuns-audience-tongue-lashing-china-russia-eve-arctic-council-ministerial.
140 Deng, 北极安全研究 [*Arctic Security Studies*], 197.
141 Ibid.

4 The Role of the Arctic in the PRC's Comprehensive National Security

The environmental and politico-economic changes occurring in the Arctic region have implications for China's national security. In 2014, the PLA's Defence Policy Research Centre at the Academy of Military Sciences stated in its annual strategic assessment of the international security environment that China, being situated in the Northern Hemisphere, has strategic interests in the region related to the sustainable development of its economy and national security. The assessment highlighted that apart from the environmental impact on China's weather system, the Arctic region is expected to become a vital future supply base for China's energy consumption, while the shorter Arctic shipping routes could influence China's future maritime trade.[1] More recently, the Chinese Arctic White Paper emphasised that as one of the continental states closest to the Arctic Circle, China is a near-Arctic state whose ecological environment and economic interests in agriculture and industry are directly impacted by the changing conditions in the Arctic.[2]

Simultaneously, under the leadership of General Secretary Xi, China's national security establishment has been progressively expanding, and the scope of national security has been continuously broadening under the banner of the Comprehensive National Security Outlook. As discussed in Chapter 2, this concept encompasses both traditional and non-traditional security domains, integrating 16 security areas such as military, political, economic, technological, resource, and ecological security. The adoption of such a comprehensive view of national security reflects the changing understanding of China's role in the world, its security environment, and the increasing number of areas that are vital to its development and security.

As such, the goal of this chapter is to analyse China's involvement in the Arctic within the context of the Comprehensive National Security Outlook. Since ecological security was discussed in the previous chapter, the subsequent sections will concentrate on the types of security that I believe the CCP can increase and benefit from the most, including military security (the PLA's activities regarding the Arctic); economic security, with a focus on resource extraction and shipping; technological security, encompassing innovation and the testing of advanced technologies; and political security, emphasising development and status.

DOI: 10.4324/9781003295112-4

Military Security: The PLA and the Arctic Region

In the Chinese context, military security refers to a state being free from external threats, such as invasion and war, with the ability to continuously maintain this condition.[3] At the same time, PLA analysts acknowledge that as a state's national security and development interests expand beyond its borders into global common spaces (such as outer space and the polar regions) it becomes necessary for that state to possess defensive and offensive military capabilities for a potential confrontation in these public domains in order to maintain the security of one's sovereign space.[4] Such viewpoints, combined with General Secretary Xi's statements that China should play a more active part in shaping the rules regarding the strategic new frontiers and his efforts to modernise the PLA into a world-class military force capable of safeguarding China's overseas interests, indicate the need for the PLA to develop capabilities to project power across regions and domains, seize the initiative and prevail in the global commons, as well as contribute internationally to boost China's prestige.[5]

While the Chinese armed forces have not been deployed to the Arctic region yet (at least according to publicly available information), it appears that recent incursions of the Chinese Navy into near-Arctic waters have become a regular occurrence as the PLAN's capabilities and operational radius continue to expand. In September 2015, following a joint Sino-Russian naval exercise, five PLAN ships (including three surface combatants, an amphibious warship, and a supply ship) entered the Bering Sea and transited US territorial waters near Alaska. These navy vessels did not violate international law as they exercised their right of innocent passage by sailing in a non-provocative and expeditious manner. Nonetheless, these actions drew attention from Washington, as it was the first time the Chinese Navy had entered US territorial waters without prior notification, coinciding with then US President Obama's visit to Alaska.[6] Some interpreted this as China signalling its operational capability to reach Alaska and the near Arctic waters.[7] Later that year, a trio of PLAN vessels visited Denmark, Finland, and Sweden, once again demonstrating China's global capabilities, willingness to enter new regions, and signalling its interest in the Arctic.[8] In July 2017, three Chinese Navy ships conducted joint naval drills with Russia in the Baltic Sea, marking the first-ever such exercise in that area.[9] During the same month, a Chinese intelligence-gathering vessel was observed in waters near Alaska, reportedly monitoring a US Terminal High Altitude Area Defence (THAAD) test.[10] In August, three Chinese naval vessels visited Helsinki, Finland, for a four-day goodwill visit.[11] In September, the PLAN conducted another naval drill with Russia, this time in the Sea of Okhotsk,[12] and finally, also in September, another trio of Chinese naval vessels visited Denmark.[13] More recently, in September 2022, a formation of Chinese and Russian naval vessels was spotted in the Bering Sea off the coast of Alaska's Kiska Island.[14] According to a Chinese military expert, the PLAN has aspirations for blue-water capabilities, making it increasingly normal for the Chinese Navy to explore areas it previously did not enter. The same expert adds that the Chinese Navy should not be limited to its own nearby waters and should extend its operations to the Arctic Ocean to understand the region and ensure navigational safety there.[15]

The PLAN's forays into these near-Arctic regions, particularly the Northern Pacific, play a crucial role in China's strategic focus on what has been previously referred to in Chinese military literature as the "Two Oceans region" (两洋地区). This region is described as an "arc-shaped strategic belt" (弧形战略地带) encompassing the Pacific and Indian Oceans, as well as the adjacent coastal areas of Asia, Africa, Oceania, North and South America, and Antarctica.[16] Given the strategic significance of this area for China's future development and security, as well as its role as a gateway not only to the Atlantic Ocean and the Mediterranean Sea but also to the Arctic Ocean, China will strive to create favourable conditions to establish itself in this Two Oceans region, participate in ocean resources extraction, and position itself to advance the development of the polar regions.[17]

Regarding the Arctic itself, PLA strategists and their civilian counterparts value the region for its geostrategic location. This is unsurprising, as historically, major powers in the international system located in the Northern Hemisphere have always been interested in the geostrategic advantages offered by the Arctic's location, particularly in terms of reducing distances between adversaries.[18] In this context, PLA analysts often refer to the region as the strategic commanding heights[19] that overlook and connect the three core economic and political areas of the Northern Hemisphere: Europe, Asia, and North America.[20] Recognising this positioning, Chinese military writings acknowledge the Arctic as a vital aviation hub or crossroads. This indicates that if an intercontinental ballistic missile were to be deployed in the Arctic, it would only require a range of 8,000 km to pose a threat to major states in North America, Europe, and Asia, significantly reducing flight distances and increasing missile defence penetration capabilities.[21] It is equally important to note that in a hypothetical nuclear exchange between the US and China, some of their intercontinental ballistic and hypersonic missiles would pass over the Arctic region. Chinese scientists recently conducted a study simulating the flight of a hypersonic missile launched from Shandong Province in eastern China to New York, with a trajectory over the Arctic region.[22] Zuo Pengfei from China's National Defence University, in his analysis of the military value of the Arctic region, highlights the notable improvements in a state's situational awareness and missile early warning capabilities by deploying long-range radars on the shores of the Arctic Ocean.[23] From the Chinese perspective, this is an area where China would greatly benefit from Russia's assistance.[24] In this regard, Hsiung points out that in recent years, Chinese and Russian experts have openly discussed missile defence cooperation, and in 2016, the two countries conducted joint computer-simulated missile defence drills.[25] At the same time, Chinese analysts argue that similar American missile defence systems located in the Arctic (in Alaska and Greenland) pose a threat to Chinese security. Lu and Zhang argue that while the US claims that these systems are part of its strategic defence network, they are, in reality, strategic pressure points aimed at Russia and China to monopolise geostrategic hegemony.[26]

In the maritime domain, Chinese analysts regard the Arctic as a strategic maritime passage that connects "three continents and two seas."[27] For example, Qu et al. observe how military great powers, such as the US and the former Soviet Union, have occasionally used the Arctic Ocean to move their submarines between the Atlantic and Pacific Oceans.[28] Others see marginal seas in the Arctic, such as the Barents Sea or the Chukchi Sea, as convenient shortcuts for maritime movements in areas where states engage in maritime struggles.[29] However, many agree that the Arctic is an ideal location for nuclear deterrence, particularly for concealing strategic nuclear submarines.[30] This is because the unique Arctic environment, characterised by sea ice and electromagnetic interferences, impacts the effectiveness of existing anti-submarine warfare technology.[31] Chinese analysts also point out that during the Cold War, countries like the Soviet Union used the Arctic Ocean as an important site for deploying submarine-launched ballistic missiles to secure second-strike capabilities. They further believe that the US and Russia still consider the region important for operating strategic nuclear capabilities today.[32] This has led some Chinese analysts to conclude that by controlling the Arctic region, one would have convenient access to three continents and two oceans, firmly establishing an advantage in military struggles.[33]

The expansion of China's global engagements and the notable rise in its naval capabilities, both qualitatively and quantitatively, have convinced some Chinese analysts that Chinese naval forces should venture into the Arctic region (as well as the seas around Antarctica) in order to protect China's national security and interests.[34] This has sparked debates among Western and Chinese analysts regarding what China stands to gain from deploying its armed forces in the region. Brady, for example, argues that the region is linked to China's nuclear security, particularly in terms of enhancing China's nuclear deterrent capabilities.[35] Zuo's analysis also focuses on Chinese nuclear security, emphasising how Arctic sea ice cover could contribute to the survivability of Chinese nuclear submarines and strengthen China's second-strike capabilities. He further suggests that if Chinese forces establish themselves in the region, they could better monitor regional US activities and provide advanced warning to the Chinese mainland in the event of incoming US missiles, thereby extending their response time and significantly reducing US nuclear pressure on China.[36] Other analyses propose that China might use its military presence in the region to defend its commercial interests, intimidate its adversaries, threaten US territorial security and those of its allies, as well as deny the region as a safe haven for the US Navy.[37] On the other hand, Lajeunesse and Choi conclude in their comprehensive analysis of the potential deployment of Chinese nuclear submarines to the Arctic Ocean, that "there is little advantage and much risk involved in extending PLAN submarine operations into the Arctic, at least as a means of threatening the United States or her allies."[38] They provide several reasons for this assessment, considering that the US and the broader North American Arctic would be the primary target of Chinese Arctic military deployments. First, the volume of trade conducted through the Northwest Passage or the Transpolar Sea Route is currently negligible, with not much to threaten. Second, the likelihood of the US using Arctic sea routes to send reinforcements to Asia in the event of a confrontation with China is low, as it would depend on various circumstances such as the availability of ice-strengthened ships and sea ice conditions in the region. Consequently, the Arctic is deemed "a poor candidate" for Chinese sea denial. Third, the chances of using the Arctic Ocean as a missile-launching position for Chinese submarines, while possible, remain low. This is primarily due to the dangers posed by the Arctic Ocean's complex hydrographic and ice conditions to Chinese vessels, as well as the high probability of detection due to American listening systems and arrays located near the Bering Strait (making it difficult for Chinese submarines to enter the region unnoticed).[39]

Considering both sides of this debate, there is still an argument to be made that the PLA will strive to venture into the region in the near future. This is primarily due to two reasons. First, the PLAN's possible future excursions into the region should be regarded as an important factor in China's ambition to establish a world-class military in the coming years. General Secretary Xi has consistently promoted and advanced this ambition over the past decade.[40] While there is no official definition of what constitutes a "world-class military," Fravel, in his study on the origins and implications of this term, concludes that it means belonging to an elite category and being among the best.[41] He also notes that Chinese military commentaries regard the US and Russia (and sometimes the UK and France) as examples of such world-class militaries.[42] What these militaries have in common is the capability to project power beyond their immediate regions, including military engagement in the Arctic. This is particularly true for the US and Russia, but the UK has also refocused its military attention on the region in recent years.[43] If China aims to be a part of this exclusive club, it needs the capacity to enter the region at its own discretion. Furthermore, if the Chinese Navy is to be a true blue-water navy and genuinely compete with the US Navy, it must undertake missions that are more ambitious and complex, and encompass larger geographic scopes,[44] which includes the Arctic region. Scholars He and Liu from the Ocean University of China argue that as a potential global power, China needs to develop a comprehensive, professional, and mobile military and logistics support force capable of operating in the environmental conditions of the Arctic Ocean.[45]

Military drills with friendly forces in sub-Arctic waters appear to be a step in that direction. Following one such drill in the Sea of Okhotsk in 2017 with the Russian Navy, Zhang Junshe, a senior research fellow at the PLA Naval Military Studies Research Institute, stated that these drills, conducted in a colder climate and at higher altitudes, enhance the flexibility of the PLAN under diverse conditions, which differ from the subtropical regions where China typically conducts naval drills.[46] Similarly, the Chinese Navy can utilise its assets in the Bohai Sea, a marginal sea enclosed by the Shandong and Liaodong Peninsulas that freezes during winter months, to conduct cold-weather training. The PLAN currently operates two Type 272 icebreakers, the Haibing 722 and Haibing 723. These 103-metre-long ships, independently designed and built in China, perform tasks such as examining sea ice conditions, ice breaking, and search and rescue in the waters of the Bohai Sea.[47] As for the PLA Army, there are locations like Mohe County in Heilongjiang province, which is the coldest and northernmost area in China with Arctic-like conditions. The Chinese Army has stationed soldiers in these areas which are, by default, training in cold-weather warfare.[48] In 2021, soldiers based in Heilongjiang province participated in the Sayan March international competition, organised by Russia, to test and enhance their winter skills in extreme and mountainous conditions.[49] According to Song Zhongping, a Chinese military analyst and former PLA instructor, this competition provided the Chinese armed forces with valuable experience, especially now that "the Chinese military's missions have expanded from safeguarding its national interests to protecting overseas interests as well, which would cover places with rough conditions and extreme weather."[50]

Second, Chinese military strategists may view the Arctic region as a highly advantageous diversionary theatre.[51] Dean and Lackenbauer argue that even a limited deployment of the Chinese Navy could trigger an overreaction from Arctic states, given their fixation on Arctic sovereignty and the intensifying strategic competition, and compel them to redirect their attention and limited resources from other strategically important regions to the Arctic.[52] Such reasoning aligns with the Chinese military concept of "exterior lines operations," originally formulated by Mao Zedong to guide the Chinese Red Army's engagement with more capable and better-equipped adversaries.[53] Rolland further explains that when facing a superior opponent, the weaker side cannot hope to defeat it solely by engaging on the interior lines. Instead, "the weaker force must engage in multiple small campaigns along exterior lines, attacking and harassing the enemy in areas where it is weaker in order to distract and disperse enemy's attention and loosen its grip on the main battlefield."[54] In the current setting of expanding US-Chinese strategic competition, one can envision a scenario in which the PLAN dispatches a submarine to the Arctic, prompting an overreaction from the US political and military establishment, leading to a redeployment of assets from the Western Pacific to the Arctic. Such a response could relieve some of the strategic pressure China faces in a region adjacent to its territorial borders.

Additionally, Lajeunesse and Choi present a third possible rationale for China's military interest in the Arctic region, which involves a hypothetical confrontation between China and Russia.[55] Despite the current cooperative stance between China and Russia aimed at challenging US dominance over the international order and their proclamation of a "no limits" partnership,[56] such scenario seems highly unlikely. Yet one cannot dismiss the possibility of deteriorating Sino-Russian relations in the future. Historically, their relationship has undergone various phases, including a close partnership in the early 1950s, enmity from the 1960s to the late 1980s, a pragmatic partnership in the 1990s and 2000s, and the current phase of potential allies.[57] In the event of a conflict between China and Russia, the deployment of Chinese forces to the Arctic region would hold significant strategic value. Russia possesses crucial hydrocarbon production facilities in its Arctic zone that are vital to its economy, and it relies on Arctic shipping routes, particularly the Northern Sea Route, to transport those resources out of the region. Chinese forces operating in the Arctic Ocean could pose a threat to these facilities and shipping routes.[58] Furthermore, Chinese naval presence in the Arctic could deny Russia the use of Arctic waters as sanctuaries for its strategic submarine forces, requiring the Russian Navy to allocate precious resources to track them. As a result, the number of assets Russia could deploy in the Pacific arena would be limited.[59] However, at present, it seems that the Chinese armed forces will have to overcome some significant obstacles before being able to pursue any of the previously mentioned options.

This is primarily due to the distinctive physical and environmental characteristics of the Arctic region and its marine domain, which require specialised knowledge, equipment, and capabilities.

Chinese Strategic Science in the Arctic

The Arctic region with its unique atmospheric, marine, and geophysical features can pose challenges for the armed forces and navies of great powers seeking to operate in this area. Factors including the presence of sea ice, extremely low temperatures, and salinity layers in the Arctic Ocean create a distinct marine environment. These conditions can provide submarines with cover, further complicating anti-submarine warfare operations.[60] Chinese scientists also point to the impact of regional climatic changes on the Arctic Ocean's hydroacoustic environment. Yin Li et al., from the Chinese Academy of Sciences, emphasises that global warming affects the layering of water beneath the Arctic sea ice. The influx of warmer water can lead to the formation of a new water layer in certain parts of the Arctic Ocean, creating a new stable acoustic channel that significantly extends the operational range of intermediate frequency sonar from 100 km to over 400 km.[61] Additionally, natural phenomena specific to polar regions, such as the aurora borealis or geomagnetic storms, can have adverse effects on radio communications, over-the-horizon radar systems, satellite signals, and the reliability of GPS. These technologies are essential for conducting military operations.[62] To develop the necessary capacities to operate in this unique environment, major military powers must first collect data through various environmental sciences to enhance their Arctic situational awareness. This concept involves acquiring a "basis for understanding the unique elements of the Arctic environment" and developing "abilities to predict its future state."[63]

The authoritative Chinese *2020 Science of Military Strategy* emphasises the importance of enhancing polar situational awareness, monitoring and communication capabilities, as well as air, sea, and land manoeuvrability in polar conditions for the Chinese armed forces.[64] Chinese analysts also highlight the importance of Arctic scientific knowledge and understanding the region's unique environment for China's ability to strategically engage in the Circumpolar North, be it military, economic, or governance-related. Deng, for instance, maintains that countries investing more in science and technology possess higher levels of what he terms knowledge power (知识性权力). This knowledge power enables these countries to expand their participation in the utilisation of polar regions and influence the shaping of regional regulations.[65] Xiao Yang, in his analysis of the strategic implications of Chinese Arctic scientific research, argues that the state that will possess the most advanced Arctic-capable technology and comprehensive Arctic information and knowledge will ultimately prevail in the great game unfolding in the region. Consequently, he advocates for China to extensively study the Arctic region to expand its knowledge base and actively participate in regional affairs.[66]

Historical precedent supports this strategic approach to Arctic engagement through science. The experiences and practices of China's principal adversary, the US, and its engagement in the Arctic region during the early years of the Cold

War, provide insights. With deteriorating relations between the US and USSR after the end of World War II and advancements in military technologies, such as long-range bombers, ballistic missiles, and later nuclear submarines, the Arctic's strategic importance increased dramatically.[67] Given its proximity to the heartlands of the US and the USSR, the region was perceived as a likely theatre for World War III.[68] However, this posed a challenge for US military planners for two reasons. First, they recognised that defending the North American continent and effectively utilising new military technologies and weapon systems necessitated a comprehensive understanding of the Arctic's physical environment, which they lacked at that time.[69] Factors like local gravity variations and ionospheric phenomena were seen as influencing the accuracy of guided missiles and communications in the northern region.[70] The US military also needed insight into the impact of snow, ice, and permafrost on military operations, including ice sheet construction, overland vehicular transportation, and unplanned landings and take-offs from ice-covered areas.[71] Second, US defence officials acknowledged that the Soviets, their superpower adversary, possessed superior knowledge of the Arctic environment compared to American scientists.[72] Consequently, the US military initiated research funding focused on, among other things, meteorology, geology, glaciology, seismology, and oceanography. They established projects across Alaska, the Canadian Arctic, and Greenland, such as weather stations in Greenland, an Arctic Research Laboratory at Point Barrow, Alaska, and a US-Canadian Joint Experimental Station at Fort Churchill in the Canadian Arctic.[73] This ongoing research in the Arctic spurred the establishment of dozens of major university research programs, training hundreds of scientists, and securing valuable data for advanced weapons systems.[74] This development of strategic Arctic science persisted throughout the Cold War, and as Pedersen's research indicates, it continues to this day. Notably, countries such as the US, Russia, but also the likes of Germany, France, the UK, and China are actively involved, gradually unveiling the region's "secrets" and levelling the Arctic battlefield.[75]

In China, the pursuit of strategic science in the service of the national interest has been further accentuated under General Secretary Xi's strong emphasis on the MCF strategy. As outlined in Chapter 2, the CCP now promotes joint construction and utilisation of civilian and military infrastructure, as well as joint exploration of the oceans, air, and outer space, aiming "to bring into place an all-element, multidomain, and cost-effective pattern of civil-military integration."[76] Recent reports indicate an active implementation of this strategy. For instance, in 2017, a Chinese research vessel conducted oceanographic surveys near Guam, the site of a major US naval base. The lead scientist from that expedition stated that all findings would be shared with interested actors within the Party-state system, including the Chinese military.[77] Concerning the sharing of physical infrastructure, the Chinese Navy has recently made high-profile visits to overseas ports controlled by Chinese state-owned enterprises, such as in Sri Lanka or Greece,[78] which could potentially serve dual-use purposes as intelligence or logistic hubs in the future.[79]

With regards to the Arctic region, Chinese analysts and publications indicate that the MCF approach is preferred for regional military engagement and the collection of relevant scientific data with dual-use applications. The 2020 edition of the SMS states that through MCF, the great powers of the international system can not only establish a military presence in the polar regions but also avoid, to some extent, international provocations and provide strong support for their own polar interests.[80] Deng argues that due to the close link between the Arctic region and US and Russian security concerns, conducting military operations in the Arctic Ocean may face significant limitations. Therefore, he suggests that in the initial stages of capacity building, China should focus on the civilian dimension of the preparatory work, such as scientific expeditions focused on Arctic Ocean geomorphology, marine acoustics, magnetism, and so forth, and incorporate the findings into the Navy's long-term planning and strategy.[81]

Because of the CCP's goal of transforming the PLA Navy into a true blue-water force capable of safeguarding the PRC's overseas interests and, if necessary, enforcing its will in distant regions, acquiring essential capabilities and relevant knowledge to operate in a contested Arctic environment is increasingly essential. Martinson and Dutton note that China's expanding naval outlook has created a demand to master the "ocean battlespace environment," including in new ocean regions.[82] Some of these demands are addressed directly by the PLAN, which, like the navies of other great powers, operates a growing fleet of distant-ocean surveying ships.[83] However, given the geopolitical sensitivities of the Arctic region and the challenges posed by its unique environment, the PLAN can leverage the civilian institutions within the Party-state system, such as Arctic science and research-focused organisations, to acquire relevant regional data, observations, and expertise.[84] As mentioned earlier, China's Arctic scientific expeditions extensively engage in oceanographic research, which includes, among other things, studies of the conjoint effects of temperature and salinity in the Arctic Ocean.[85] These efforts yield crucial data that can have dual-use applications. Beyond the marine environment, in Iceland, China has been operating an Arctic Observatory in cooperation with Icelandic research organisations since 2018. The observatory's research focus includes the study of phenomena like the aurora borealis, space weather, meteorology, climatology, and more.[86] Qu et al. point out that changes in these atmospheric phenomena directly impact the operation and reliability of space and ground military and civilian technical systems, including satellite failures, communication disruptions, navigation and tracking errors, power system outages, and so on. Under such circumstances, the normal functioning of military command and control systems, space-based military equipment, and precision-strike weapons is affected.[87] Thus, establishing an Arctic multidimensional research and monitoring system and having the ability to accurately predict atmospheric variations in the region would provide strategic, operational, and tactical advantages for any country, including China, with interests in the Arctic.[88]

Given China's strong emphasis on MCF, one can reasonably expect that the data generated during Chinese scientific expeditions, whether focused on oceanographic or terrestrial research, can be shared across the Party-state system. The National Marine Environment Forecasting Centre, a civilian agency under the MNR, regularly supplies oceanographic and meteorological products and data to the PLAN.[89] Notably, the centre also operates the Polar Environment Research Forecast Office, which specialises in developing and producing relevant forecast data related to polar oceans, sea ice, and meteorology.[90] Similarly, Martinson and Dutton highlight the involvement of the Third Institute of Oceanography under the MNR in military oceanography and national defence construction projects.[91] This Institute also actively participates in Chinese polar research, engaging in research projects and data acquisition related to the Arctic Ocean.[92]

Additionally, from a strategic standpoint, the geographical scope of Chinese Arctic scientific expeditions may lead to concerns that they are designed to gather information to facilitate Chinese naval operations in the sub-Arctic waters of the North Pacific or the Arctic Ocean itself.[93] The primary areas of operation and data collection for Chinese Arctic scientific expeditions are the Bering, Chukchi, and Beaufort Seas. Although these areas are relatively close to the Chinese mainland and serve as entry points to the Arctic Ocean for both civilian or military vessels, Millard and Lackenbauer argue that they are crucial territories, particularly the Bering and its surrounding seas, akin to the strategic importance of the Greenland-Iceland-UK (GIUK) gap in the sub-Arctic waters of the North Atlantic. From a Chinese perspective, it would be strategically unwise not to thoroughly survey these areas.[94] Considering that the Arctic Ocean will become increasingly accessible in the near future and more integrated into the global maritime system, whether due to the emergence of new shipping lanes or heightened regional great power competition, the GIUK gap is likely to become an important and contested gateway.[95] A similar dynamic will likely apply to the sub-Arctic waters of the North Pacific. When it comes to the potential involvement of strategic Arctic science, the case of China's 10th Arctic scientific expeditions provides insights. Launched in August 2019, this expedition conducted multidisciplinary investigations and surveys in fields such as marine chemistry, biology, ecology, geology, geophysics, and physical oceanography in the Bering Sea, Bering Strait, and Chukchi Sea.[96] Notably, the expedition took place aboard the advanced oceanographic research vessel Xiang Yang Hong 1,[97] operated by the First Institute of Oceanography under the MNR. The vessel, constructed in 2016, is specifically designed to serve as a platform for collecting oceanic and atmospheric data, deploying surface and subsurface buoys, and accommodating UUVs.[98] In early 2021, naval observers expressed interest in the Xiang Yang Hong 1 and similar research vessels when it was observed conducting research, likely involving seabed mapping, in the eastern Indian Ocean – an area "likely to be of particular interest to the Chinese Navy as they expand their submarine capabilities."[99] Although there are no available sources confirming whether this research was conducted on behalf of the PLAN, either in the Indian or Arctic Oceans, the data produced are "civilian-defence agnostic," meaning they can be used for both civilian and military purposes.[100]

Within the scope of the MCF, various other Chinese civilian-led projects established in the Arctic region can be used by the PLA to expand its global reach. One notable example is the development of ground segments, including ground receiving stations, which support aspects of the PRC's space program, including the Beidou navigation satellite system.[101] Beidou is a global navigation system independently developed by China to serve its national security, economic, and social development needs. It currently provides all-weather and high-accuracy positioning and navigation services to users worldwide.[102] In addition to its civilian applications, Beidou also assists the Chinese military and China's maritime militia in contested areas like the South China Sea.[103] Beidou itself is a product of China's MCF strategy, and according to Strickland, the Party-state effectively leveraged numerous academic institutions, including Wuhan University, to facilitate its creation and refinement.[104] Notably, specialists from Wuhan University aided China's State Bureau of Surveying and Mapping in installing a Beidou satellite navigation and positioning reference station at China's Yellow River Station in Svalbard, Norway, in 2016. The Chinese State Bureau of Surveying and Mapping stated that this installation would not only enhance China's polar research, mapping, and navigation service capabilities but also boost Beidou's global precision capabilities, improve its position accuracy services, and accelerate its internationalisation.[105]

Furthermore, as described in Chapter 3, China successfully established a remote sensing satellite ground station in the Swedish Arctic in 2016. Operated by China's CAS, this station plays a vital role in enhancing satellite data transmission efficiency for the Chinese Gaofen project, a satellite network that provides the Party-state with global surveillance capabilities.[106] Additionally, because of the rigid Arctic conditions that put additional demands on technology and construction methods, Chinese technicians acquired valuable knowledge and expertise when constructing the ground station. Such competencies can be transferred to other sectors of China's economy or applied to similar projects in the region, thus advancing China's independent engineering capabilities.[107] When combined with China's space technology development and guided by the MCF strategy, these technological assets expand the PLA's global reach both vertically into space and horizontally well beyond China's immediate vicinity.[108] They also allow China to rely on its own hardware during polar scientific operations, an important dimension of China's technological advancement in the New Era. However, as more Western states reassess their relations with China due to the growing difficulty in clearly distinguishing between China's civilian and military domains, scientific collaborations face heightened scrutiny. Consequently, some Chinese projects, including those in the Arctic, have been affected. In 2019, the Swedish Defence Research Agency cautioned that China might be using its Arctic ground satellite stations to supplement its military intelligence efforts.[109] Subsequently, the Swedish Space Corporation, Sweden's state-owned space company that holds the contracts allowing China access to the antennas in Sweden, announced that it would not renew those contracts.[110]

However, this does not imply that China will be completely excluded from accessing ground facilities in the Arctic. For instance, in 2017, Beijing Normal University expressed interest in establishing a ground satellite station in Greenland.[111] Additionally, China is actively integrating and interoperating its navigation and positioning systems with those operated by Russia. In 2022, the two countries signed an agreement for the construction and operation of Beidou and Glonass tracking stations on their respective territories, along with a statement on jointly providing user information support services.[112] Although this agreement is not specifically focused on the Arctic, the possibility of closer cooperation between these two aligned states, including the construction of ground segments of their space-based systems in the Arctic, should not be overlooked.

Moving beyond strategic science support for PLA's Arctic situational awareness, another MCF approach that the CCP might adopt in the coming years, as the PLA's long-distance power projection capabilities improve, is the direct involvement of the Chinese armed forces in the logistical, technological, and medical support of Chinese Arctic scientific expeditions. China has previously undertaken such a mission in the 1980s. In 1984, at the direction of the CCP leadership, the PLAN and the SOA jointly organised China's first Antarctic expedition, with the primary objective of establishing an Antarctic research station on King George Island near the Antarctic peninsula.[113] The military provided substantial organisational, logistical, and personnel support for the expedition.[114] It involved two vessels, the Xiang Yang Hong 10, an oceanographic research ship, and the J121, a naval salvage lifting ship, as well as 591 participants, half of whom were military personnel.[115] The PLAN admiralty at the time believed that the expedition would contribute to training Chinese military personnel in distant-seas navigation and thus have a long-term impact and strategic significance.[116]

Present-day Chinese analyses point to such military logistical approach by other great powers towards the polar regions, including the Arctic, and suggest that these could serve as examples for the Chinese armed forces to emulate. The 2020 SMS notes that the military forces of the US, Russia, the UK, France, Japan, and South Korea actively participate in supporting their polar scientific efforts.[117] Furthermore, the technological and logistic support provided not only contributes to the military's own scientific research but also offers opportunities for these armed forces to gain experience in cold-weather operations.[118] Yin Li et al., in their analysis of Chinese and international research on Arctic underwater signal processing, provide an example of the US Navy's involvement, particularly its nuclear submarines, through the Submarine Arctic Science Program – a collaboration scheme involving multiple agencies – to conduct scientific studies in the Arctic Ocean, highlighting its civil-military nature.[119] Song Yunxia from the Dalian Naval Academy suggests that such military-in-support-of-science missions, which she labels as covert military operations (隐性军事活动), where military forces are engaged but led or funded by other agencies with non-military objectives, should be one of the primary forms of Chinese military operations in the polar regions.[120] The 2020 SMS emphasises in its strategic guidance for polar military force application that military and civilian strengths should be closely integrated, and that it is essential to fully utilise the military's role in supporting polar scientific research.[121] This support could take the form of technical and medical equipment, as well as logistical assistance for national polar scientific expeditions, including the use of large transport aircraft, military support vessels, and specialised ground transport vehicles.[122]

Economic Security: Resources and Shipping

While considered the foundation of China's national security,[123] economic security primarily relates to efforts in maintaining the basic socialist economic system in the country. Its core objectives include continuous economic development, enhancing overall economic strength, and competitiveness, the ability to withstand internal and external shocks and threats, and a focus on preventing and controlling major risks and challenges that could hinder China's development.[124] Energy is closely linked to economic security and the overall societal advancements of the Chinese nation. In recent years, with the changing global energy landscape and China's emphasis on green development, concerns, and discussion regarding energy security, including sources, supply, and affordability, have become integral to high-level CCP meetings and conferences.[125]

Currently, China holds the position of the world's largest energy consumer and the leading importer of crude oil[126] and liquified natural gas.[127] While the Party-state leadership has pledged to reshape China's energy mix, aiming to reduce CO_2 emissions and achieve carbon neutrality by 2060,[128] and the 14th Five-Year Development Plan emphasises renewable energy technologies, nuclear power, and intelligent energy solutions,[129] short-term projections indicate an expected increase in China's energy consumption. Sinopec, for instance, in its recent energy outlook anticipates Chinese energy consumption to peak by the mid-2030s.[130]

Despite the CCP's desire to produce more energy within China to mitigate the impacts of geopolitical challenges and volatile oil prices,[131] the country remains heavily reliant on overseas imports (approximately 50% of oil imports originate from the Middle East).[132] Additionally, a significant portion of these energy imports passes through straits and transportation bottlenecks near the contested waters of the South China Sea, which could be vulnerable in the event of a Sino-American naval confrontation. Consequently, Chinese analysts deem it prudent and necessary to diversify China's energy import sources and supply routes to ensure energy security and long-term development prospects.[133] Some experts argue that the energy potential in the Arctic could contribute, to some extent, to such diversification and energy security. For instance, Sun and Hao from the Ocean University of China in Qingdao note that Arctic energy development is related to the stability and sustainability of China's national strategic development.[134]

The Arctic region, with its abundant natural resources, especially oil and gas, is occasionally compared to other resource-rich regions such as the Middle East in China's state-linked media, which the country should consider in its long-term strategic development.[135] References to the Arctic's resource abundance can also be found in Chinese official documents, academic analyses, and discourses. For instance, the PRC's 12th Five-Year Development Plan for Polar Expeditions (2011–2015) notes that "the Antarctic and Arctic regions are rich in resources."[136] The Chinese Arctic White Paper includes a section on the utilisation of Arctic natural resources, affirming that "the Arctic has abundant resources," while encouraging "its enterprises to engage in international cooperation on the exploration for and utilisation of Arctic resources."[137] In academia, Lu and Zhang maintain that the Arctic is a globally significant treasure trove of natural resources, destined to become an irreplaceable energy supply source in the future.[138] Similarly, a 2022 article in the state-run Xinhua News Agency's general affairs magazine, *Outlook* (瞭望) compares the Arctic's resource potential to the Middle East, projecting it to become an important energy-producing area in the second half of the century.[139]

Additionally, the Chinese Party-state recognises the clean-energy potential of certain Arctic regions, including wind, solar, and geothermal energy resources.[140] Gao et al., in a recent publication, advocate for the development of the renewable energy potential in the Russian Arctic and the Far East, emphasising abundant hydropower, geothermal, wind, solar, biofuels, and ocean energy resources. They argue that such potential represents an investment opportunity for Chinese companies, further diversifying Sino-Russian energy cooperation.[141]

Beyond energy resources, Chinese state-linked actors are also drawn to the mineral potential of the Arctic, including diamonds, uranium, nickel, copper, gold, and rare-earth elements, and their availability due to environmental changes in the region.[142] Greenland, known for its vast rare-earth deposits, has been a focal point of Chinese mining ambitions in the Arctic. Researchers from the Chinese Academy of Geological Sciences highlight Greenland's favourable rare-earth resource availability, deserving attention from Chinese enterprises.[143] China's SOEs, such as Shenghe Resources, have previously been involved in Greenland, with stakes in the Kvanefjeld site in the south of Greenland, which is rich in rare-earth elements and uranium. If developed, this project would further solidify China's dominance in rare-earth metal extraction and processing.[144] Although the project is currently stalled due to domestic Greenlandic politics (the ruling coalition had environmental concerns) and geopolitical pressures from the US, the situation may change in the long term, as there are political parties in Greenland that support the project's advancement.[145]

The stated description of some of the Chinese perspectives on the energy and resource potential of the Arctic may suggest that China's state and sub-state actors are preparing to venture into the Arctic, fuelling the narrative of the so-called "scramble for the Arctic."[146] However, it is important to note that while there is optimism in China, particularly within academic circles, about the future potential of Arctic resource development, Chinese stakeholders are fully aware of the challenges and controversies associated with it. For example, an article on the website of the MNR analyses the status of oil and gas exploration and development in the Arctic. While it concluded that China should include Arctic resources in its overseas oil and gas development strategy, the article also highlights the need for Chinese enterprises to consider the high costs and technical challenges of Arctic projects, the lack of supporting infrastructure, environmental concerns, controversies involving NGOs and Indigenous populations, as well as growing geopolitical uncertainties.[147]

So far, the most successful Chinese investments in Arctic energy have been LNG projects in the Russian Arctic zone. Currently, Chinese SOEs and investment funds control 29.9% of the Yamal LNG project (through CNPC and the Silk Road Fund) and 20% of the Arctic LNG 2 project (through CNPC and China National Offshore Oil Corporation CNOOC). Chinese development banks, such as the Export-Import Bank of China and the China Development Bank, have provided substantial credit lines for both projects, totalling around USD 14.5 billion.[148] Yamal LNG, which is now fully operational, is the world's largest of its kind and the first large-scale overseas project implemented under the BRI. It has an annual export capacity of 16.5 million tons of LNG, with 4 million tons designated for the Chinese market.

Chinese state-run media immediately lauded the project as a model for international energy cooperation in the Arctic, referring to it as an "energy pearl on the Arctic Circle" and emphasising the invaluable experience it provides for China in the development of Arctic natural resources.[149] Arctic LNG 2 is expected to be fully operational by 2026, with an annual production capacity of nearly 20 million tons of LNG, 30% of which has been contracted for the Chinese market.[150] While there have been reports that some Chinese energy companies halted or were asked to halt certain activities in Russia due to international sanctions imposed on Russia following its invasion of Ukraine in the spring of 2022,[151] this did not affect Chinese investments in the Russian Arctic LNG projects. In fact, in 2022, China set a record for LNG imports from the Russian Arctic and Far Eastern regions, spending USD 68 billion on energy imports from Russia, a 50% increase compared to 2021.[152]

As the CCP focuses on addressing air pollution, environmental protection, and climate change, one of the measures taken is the adjustment of China's energy consumption mix. Chinese analysts recognise natural gas as an effective tool in addressing these issues, as it emits less CO_2 compared to coal. Consequently, they predict China's natural gas consumption will continue to grow in the near future, highlighting the importance of ensuring a steady supply for the country's national development.[153] Russian Arctic LNG can contribute to the steady supply. According to a representative from CNPC, natural gas from Yamal will be of great significance to China in its efforts to accelerate energy structure adjustments and ensure energy security.[154]

The utilisation of Arctic shipping routes is closely related to Chinese economic security. As one of the world's largest trading nations, China heavily relies on sea transportation for its international trade (around 90% is transported by sea). This dependency creates security concerns for the Party-state, which requires sustained access to global shipping routes to fuel its economic growth. Consequently, the emergence of Arctic shipping routes has generated optimism in China. Chinese analysts believe that utilising these routes can enhance China's economic security by reducing distances and costs between Chinese and overseas ports (particularly in Europe), bypassing unsafe regions and congested transportation chokepoints, and serving as alternative energy and supply channels for the Chinese market.[155] Due to their significant commercial and strategic value, the Arctic shipping routes are often referred to as the golden waterways (黄金水道) in Chinese discourse.[156] The Chinese government also echoes some of these sentiments, primarily driven by the perceived economic potential of these shipping routes.[157] Zhang Ming, a former Chinese Vice Foreign Minister, highlights that "China pays close attention to the potential major impact of those sea routes on global shipping and trade."[158] Official Chinese documents envision the Arctic shipping routes to Europe as a blue economic passage that can foster sustainable development and economic growth.[159] The Arctic White Paper calls for the establishment of a Polar Silk Road (PSR) through the development of Arctic shipping routes "to pave the way for their commercial and regularised operation."[160] The concept of the PSR, as a third component of the BRI, alongside the Silk Road Economic Belt and the 21st Century Maritime Silk Road, is now firmly ingrained in Chinese discourse and narratives regarding the future of the Arctic region. Some Chinese analysts believe that constructing the

PSR will inject new vitality into China's participation in Arctic resource development and its engagement in regional affairs.[161]

While the Chinese government has launched scientific expeditions to the Arctic Ocean to conduct research on the shipping potential of Arctic sea routes, including the Transpolar Sea Route,[162] currently, only the NSR is considered promising for developing regularised shipping operations.[163] To support the development of such operations and the PSR, China's largest shipping company, COSCO, has engaged in international shipping on the NSR and has been promoting this alternative route to its global customers.[164] However, international experts agree that developing a safe and reliable international container shipping route connecting the Pacific and Atlantic oceans to rival traffic through the Suez Canal remains implausible due to safety concerns, lack of infrastructure, environmental uncertainties, and a short navigational season.[165] Nonetheless, Gunnarsson and Moe argue that these barriers may not be absolute. They propose that multi-purpose vessels capable of transporting various types of cargo, such as high-value bulk and time-sensitive project cargo (e.g., machinery), which COSCO has transported in the past, could be a promising scenario for shipping on the NSR in the near future.[166]

However, to fully engage with the Arctic region and maximise its economic potential, China and other states must overcome several challenges beyond infrastructure needs. These challenges include the lack of real-time information about Arctic sea ice and weather conditions, as well as potential legal and environmental obstacles associated with resource extraction. Chinese Arctic scientific research, as described in Chapter 3, is indispensable in this process. Chinese oceanographic expeditions to the Arctic routinely incorporate research tasks directly relevant to commercial shipping, such as marine hydrology, weather forecasting, and sea ice distribution prediction. During China's eight Arctic expedition in 2017, which covered the Transpolar Sea Route and the Northwest Passage, Chinese scientists aboard the research vessel Xuelong conducted seabed topographic surveys, collected meteorological and sea ice-related data, and gathered marine environmental data to promote the commercial use of these routes by Chinese companies.[167] Researchers at the National Marine Environmental Forecasting Centre are developing models to better predict Arctic sea ice concentrations, which can benefit commercial activities.[168] As mentioned earlier, Chinese research institutes also launch satellites that aid the PRC in observing the Arctic region, its climate, and its environment. The data and experiences accumulated during scientific expeditions and observations, as well as those from SOEs, can be utilised by Chinese government organisations. For example, the Ministry of Transport compiles some of this data to publish Chinese language guidelines on Arctic shipping. In recent years, guidelines for the Northwest Passage were published in 2015,[169] followed by the Atlas for Arctic navigation.[170] In 2017, the Arctic Northeast Passage Communication guidelines were released,[171] and in 2022, updated guidelines for the Northeast Passage (which includes the NSR) were issued.[172] Additionally, Party-state organisations, as outlined in Chapter 3, conduct research relevant to natural resource extraction in the Arctic. For example, units under the CGS have previously published reports on the hydrocarbon resource potential of the Arctic Ocean and its adjacent seas.[173]

Technological Security: Innovation and Testing of
Advanced Technologies

Within the Chinese Party-state system, technological security refers to the integration and effectiveness of a science and technology system that ensures the security and controllability of core technologies in key national areas and which has the ability to protect national interests and security from external technological advantages.[174] Advanced scientific and technological capabilities are viewed as crucial factors in shaping the balance of political and economic forces globally, enhancing a state's international competitiveness, overall national strength, and ensuring national security. In this regard, technological security is closely intertwined with the military and economic aspects of national security, making it vital for safeguarding China's national security. However, the dominance of Western states in certain technological fields and the competition for science and technology talent are identified as notable long-term challenges to China's technological security.[175]

Through technological innovation and applications developed for operations in the Arctic region, China could alleviate, to a certain degree, some of these long-term challenges. The Yamal LNG project serves as a notable example. While the development of this project provided substantial business opportunities for Chinese companies (45 such companies were involved in servicing the project),[176] it also offered valuable experience in Arctic exploration, transportation, planning, technological innovation, and production, as reported by the state-run Xinhua News Agency.[177] CNPC, one of the main investors, announced that through this project, the company gained valuable knowledge about cold-weather engineering, laying the groundwork for further expansion of Sino-Russian Arctic energy cooperation.[178] Additionally, the Offshore Oil Engineering Company, a subsidiary of the powerful SOE CNOOC, signed a USD 1.6 billion contract to construct and deliver core modules for the liquefication process on the Yamal LNG project.[179]

Building on the success of Yamal LNG and the significant technological and financial contribution from Chinese SOEs, a substantial share of technology contracts for the development of the Arctic LNG 2 project were awarded to Chinese companies. According to Spivak and Gabuev, Penglai Jutal Offshore Engineering, BOMESC Offshore Engineering, McDermott Wuchuan (a joint venture between McDermott and China State Shipbuilding Corporation), all of which participated in Yamal, as well as Wison Offshore and Marine, will manufacture generators, compressors, and modules worth hundreds of millions of USD for the project.[180] Reports also indicate that another Chinese company, the Honghua group, specialising in manufacturing drilling equipment, successfully developed, tested, and fabricated an oil drilling rig capable of functioning in extremely cold conditions. In 2019, it was delivered to a Russian customer for operations above the Arctic Circle.[181] Other SOEs, such as China National Chemical Engineering, agreed to deliver oil processing equipment to the Payakha Oilfield in the Russian Arctic.[182] The knowledge and experience gained by Chinese actors through their participation in Arctic projects can be transferred and applied to other energy development projects in the near future.

Another technological area where China has the potential to achieve break-throughs through its increased involvement in the Arctic region is the shipbuild-ing sector, particularly in the domestic development and construction of Chinese ice-class ships. The successful construction of China's second research icebreaker, the Xuelong 2, at the Jiangnan Shipyards in Shanghai, and the assembly of several Arc4 ice-class LNG tankers for the Yamal LNG project at the Hudong-Zhonghua shipyards (both subsidiaries of CSSC, China's leading corporation in naval ves-sel research, design, production, and testing) are examples of this progress. More recently, in 2021, the Chinese Ministry of Transport announced plans to develop a new icebreaker for operations along the Polar Silk Road.[183] Within the next three to five years, they aim to have the preliminary design scheme for a heavy-duty icebreaking salvage and rescue ship in order to enhance China's emergency ca-pacities along the PSR.[184] Although the notice did not specify where exactly this new heavy-duty icebreaking ship would be constructed, it is reasonable to expect it to take place at the Jiangnan shipyards given their previous experiences with constructing the Xuelong 2.

Beyond conventional icebreakers, Chinese SOEs are also exploring the pos-sibilities of domestically building a nuclear-powered icebreaker. In early 2018, the China National Nuclear Corporation (CNNC) made headlines by announcing that it was accepting bids from domestic companies to build China's first nuclear-powered icebreaker.[185] The idea had been promoted by the corporation since at least 2014[186] and was included in the State Council's 13th Five-Year National De-velopment Plan for Strategic Emerging Industries.[187] The contract was awarded to Shanghai Jiaotong University, which later established the Research Institute for Nuclear-Powered Ships and Maritime Equipment in Shanghai in collaboration with CNNC and the Shanghai Nuclear Power Office in order to focus on the devel-opment of related marine nuclear technology.[188] The press release from the opening of this research institute highlighted cooperation on projects such as the nuclear-powered icebreaker as a typical example of military-civil fusion.[189] Chinese re-ports also note that the development of this technology will enhance China's polar research capabilities and benefit its small reactor technology, particularly floating nuclear power plant technology, and the development of large-scale nuclear-pow-ered surface warships such as aircraft carriers.[190] This is because, according to the US Department of Defence, China's state-owned enterprises frequently exchange information on ship design and construction to improve shipbuilding efficiency.[191]

Chinese engagement in Arctic scientific research also requires the development of specialised equipment, as conducting oceanographic surveys and collecting data in the harsh Arctic environment demands advanced technologies.[192] Chinese ana-lysts and sources emphasise the significance of the Arctic in advancing such tech-nologies, many of which have military or dual-use applications. The 2020 edition of the SMS highlights that the polar regions, with their extreme physical and elec-tromagnetic environments naturally serve as testing grounds for various materials and equipment.[193] Dong Yongzai from the Chinese Academy of Military Sciences similarly recognises the unique Arctic conditions as valuable for testing cutting-edge military technologies under extreme conditions and as an important training space for military personnel.[194] Song suggests China should deploy its own military specialists and equipment to its research stations in the polar regions to enhance its overall technological capabilities, drawing from American and Russian experiences and practices.[195]

The testing of various UUVs in Arctic and sub-Arctic waters exemplifies the development of Chinese technologies with dual-use purposes. As described in Chapter 3, Chinese research institutes have deployed numerous UUVs in the Arctic region since 2008, such as the Haiyi (海翼) underwater glider during the ninth Arctic scientific expedition and the Tansuo 4500 UUV during the 12th Arctic scientific expedition. These UUVs are used to gather diverse marine data, including ocean topography, water salinity, and ice thickness and movement. While civilian actors can use this data to study Arctic environmental changes or develop commercial shipping routes, military analysts note that these unmanned systems and their data-collecting capabilities can also enhance the PLA Navy's tactical understanding of the ocean environment.[196] Additionally, Liu and Ma from the PLA's Naval Research Academy in their analysis of the application of this technology highlight that UUVs can be employed in search-and-rescue operations, intelligence gathering, long-term reconnaissance, and as weapons platforms for naval blockades or control of strategic sea lines of communications. Consequently, UUVs represent a technology of significant importance in future warfare.[197] The Haiyi underwater glider has already been deployed on various research missions in the contested waters of the South China Sea, providing China with methods that could strengthen its maritime surveillance and monitoring capabilities vis-à-vis its regional adversaries in these strategically important waters.[198] Similarly, Liu Zhenglu from the PLA National Defence University argues that unmanned systems hold the key to the future of Arctic military operations. Hence, extra-regional states with interests in the Arctic should make appropriate technological and equipment preparations to safeguard their interests.[199]

Closely related to the development of marine technology platforms is the study of underwater acoustical signal processing, a technology with various applications including acoustic communications, ocean floor profiling, and the detection and localisation of surface and subsurface objects, which is relevant in submarine warfare. China began conducting Arctic acoustic research in 2014, and since then, its Arctic scientific expeditions have included research tasks related to acoustic tests in the Arctic Ocean.[200] For example, during the ninth Arctic scientific expedition, Chinese scientists conducted comprehensive acoustic observation experiments in the Arctic Ocean and its adjacent waters, investigating, among other things, sound source location and underwater acoustic communication.[201]

Some Chinese analysts maintain that the PRC should further leverage its military-civil fusion strategy to expedite the development of unmanned systems and associated technologies.[202] Chinese universities are positioned well to lead these efforts. For instance, HEU manages the National Defence Key Laboratory of Underwater Acoustic Technology, which engages in a wide range of areas such as underwater acoustic physics, underwater acoustic target detection and positioning, and underwater communication technology research.[203] In 2021, the University, in collaboration with other Chinese institutions, including the PRIC, received funding from the National Natural Science Foundation of China (one of the country's major funders of basic research) to develop, among other things, "a method for predicting the vertical icebreaking ability of underwater vehicles."[204] To further enhance its expertise in polar technology development, particularly in acoustics, HEU cooperates with Russian institutions, conducting experiments and organising joint symposia.[205] Additionally, HEU, in cooperation with the Dalian University of Technology and COSCO, conducted a research project focusing on polar ship design and enhancing the navigation safety of polar class ships.[206]

These efforts align closely with the CCP's core domestic objectives, including economic restructuring, and support its long-term vision of ensuring China's role as a leading global great power.[207] By investing in and developing technological solutions for Arctic-related projects, the CCP advances the plans outlined in its industrial policies, such as *Made in China 2025*,[208] and supports the development of a modern industrial cluster in China. Jin Zhuanglong, the Minister and Party Secretary of MIIT, envisions the cultivation of national manufacturing innovation centres and the acceleration of emerging industries, including green and low-carbon industries, as part of this cluster.[209] Furthermore, equally important and relevant to China's technological security, is that Chinese Arctic technological solutions and innovations support the CCP's drive for China's technological self-reliance as stipulated in its *Dual Circulation* strategy. Some Chinese analysts believe that Chinese companies' involvement in Arctic projects and the development of associated cold-weather resource-extraction technologies have allowed them to break the foreign technological monopoly in these areas.[210] Such breakthroughs indicate that the future development of the Arctic region may not solely rely on technological solutions from a single techno-political bloc such as the West. According to Sørensen, the great power or cluster of powers that possess the best knowledge about the region and which can generate the most advanced and efficient technology suited for polar conditions will gain a strategic advantage in the 21st-century great power competition, and the CCP aims to ensure that China has that advantage.[211]

Political Security: Socio-Economic Development, International Status, and Prestige

One of the fundamental aspects of China's comprehensive national security is the importance attributed to political security, which affects all other aspects of China's security as it is closely linked to the security of the Party and the state. Internally, political security entails adherence to the leadership of the CCP and the preservation of the socialist political system, as well as the stability of the social and political order within China. Externally, it focuses on safeguarding the country's sovereignty, independence, and territorial integrity.[212] As such, at its core, political security encompasses regime and system security.[213] The CCP leadership needs to maintain its legitimacy to govern mainland China by continuously striving to deliver economic prosperity, growth, and stability, as well as meeting domestic expectations regarding China's international status and prestige.[214]

China's engagement in the Arctic region can contribute to its socio-economic development and enhance its international status and prestige. This, in turn, can help maintain the CCP's regime and ensure political security. For instance, the economic opportunities presented by the opening of the Arctic region could contribute to the economic revitalisation of Chinese provinces, especially those located in the northeast of the country. The northeast provinces of Heilongjiang, Jilin, and Liaoning are lagging in terms of economic development compared to the southern coastal provinces. Historically reliant on heavy industry, these provinces had to downsize their factories as China's economy modernised and progressively shifted towards services in the 1990s. This had a significant impact on their overall economic standing.[215] According to some calculations, the combined economic output of the Chinese northeast fell to 4.6% in the first half of 2021, down from 11.1% in 2003.[216] The CCP leadership acknowledges this ongoing problem and has taken steps to address it. In 2003, a leading small group for the revitalisation of the old industrial base in the northeast region was established, chaired by the State Council Premier and comprising various ministries and commissions, including foreign affairs, commerce, and transport.[217]

In addition, integrating the north-eastern provinces with national-level strategies, such as the BRI, can contribute to the development of their stagnant economies. The *Vision and Actions on Jointly Building the Silk Road Economic Belt and 21st Century Maritime Silk Road*, a policy plan published by the State Council in 2015, specifically mentions the north-eastern provinces and encourages their connections with Russia's Far Eastern regions to the north.[218] This has created expectations in these provinces that through the BRI and associated regional economic integration, cross-border infrastructure development in cooperation with Russia, and the establishment of new industrial and trade zones, their local economies could be revitalised.[219] Additionally, with the Chinese government strengthening the maritime component of the BRI and envisioning a "blue economic passage" through the Arctic Ocean to Europe, the north-eastern provinces started planning how to leverage the economic potential of the Arctic, particularly in terms of transportation.[220] For example, Liaoning province views Arctic shipping routes as economically viable, and ports in the province, such as Dalian and Yingkou, have been used as departure points for shipments to Europe via the NSR.[221] Similarly, landlocked provinces Jilin and Heilongjiang are involved in the construction of transportation corridors, such as Primorye 1 and 2, through cooperation with Russia, to gain access to the Sea of Japan and expand their export route options, including through the Arctic Ocean.[222] On land, Heilongjiang province sees the potential for developing connections with the Russian Far Eastern and Arctic regions as it would like to establish links with Amur Oblast and the Sakha Republic.[223] If successful, the north-eastern provinces could enhance their positions as international logistic centres with growth in the shipping, port development, and e-commerce sectors,[224] thereby addressing some of the economic challenges facing the Chinese north-east.

Moreover, as indicated previously, notions of Chinese political security are closely tied to status and prestige. Shortly after assuming the position of General Secretary of the CCP, perhaps to set a tone for his leadership, Xi Jinping, along with the entire Politburo Standing Committee, visited the National Museum of China in Beijing. There, the core of the CCP leadership toured the exhibition *The Road toward Rejuvenation*, which chronicles China's modern history since the first Opium War in 1839. Visitors to this permanent exhibition are reminded of the struggles faced by the Chinese people in the past, including foreign imperialism, but also encounter stories of the development of China's space program, the construction of the massive Three Gorges Dam, and references to China's presence in the polar regions. These serve as powerful reminders of the status and prestige that the PRC has achieved under the leadership of the CCP. The pursuit of status, defined as the collective beliefs about a state's ranking in attributes such as wealth, military capabilities, culture, and diplomatic influence,[225] seems to be a consistent behaviour among states in the international system, particularly among aspiring or rising powers. For example, Larson and Shevchenko, in their analysis of Chinese and Russian foreign policies in the post-Cold War era, identify a common objective: the desire to restore their great power status.[226] In order to achieve this, states seeking status will pursue the acquisition of status symbols in the hope of influencing other states' perceptions of their relative standing.[227] Ross observes that Chinese leaders have bolstered their status and prestige through high-profile programs (status symbols), including the development of Chinese passenger airplanes, the high-speed railway network, Lunar and Mars exploration programs, and the hosting of the Olympic games.[228] Pu and Schweller argue that these programs "are intended as costly signals of great power status, for they require enormous capabilities and resources that most countries do not possess. If such projects were to become normal and widespread state behaviours, they would no longer confer status."[229]

China's engagement with the Arctic region can also be perceived as a pursuit of status, particularly among non-Arctic states, as it requires considerable financial resources to acquire or develop the capabilities necessary for an extra-regional state to establish a presence in the Arctic. Tonami makes this point when arguing that states without a long history of polar exploration or the geographical advantage of being located in a polar region, but which are able to dedicate scarce resources to polar research and scientific programs, effectively display their symbolic power.[230] In the case of China, as outlined in Chapter 3, the country can boast two research icebreakers, two research stations, various unmanned systems, polar observing satellites, and a growing number of domestic research institutes and universities engaged in Arctic science. Li and Zhang from the Ocean University of China add that China's polar technological capabilities clearly demonstrate the strength of Chinese equipment and the "Made in China" brand.[231] Beyond the scientific and technological realm, Lajeunesse and Choi note that a potential Chinese naval deployment to the Arctic would send a powerful message to the world and the Chinese domestic audience, affirming that China, under the leadership of the CCP, is indeed a world-class technological power capable of undertaking the most challenging global deployments.[232] This could further solidify the legitimacy of the CCP in China and satisfy the growing nationalistic aspirations for international prestige among the Chinese population.

Notes

1   Tao Shelan, "军方智库：参与北极事务及其开发具有长远意义" [Military Think Tank: Participating in Arctic Affairs and Its Development Has Long-Term Significance], 中国新闻网 [*China News*], June 18, 2014, www.chinanews.com.cn/mil/2014/06-18/6292999.shtml.
2   "Full Text: China's Arctic Policy," *Xinhua*, January 26, 2018, www.xinhuanet.com/english/2018-01/26/c_136926498.htm.
3   "总体国家安全观的16种安全" [16 Kinds of Security of the Comprehensive National Security Outlook], 国安宣工作室 [*National Security Propaganda Office*], April 14, 2021, www.stdaily.com/cehua/20210414/2021-04/14/content_1114342.shtml.
4   Shou Xiaosong, ed., 战略学 [*The Science of Military Strategy*] (Beijing: 军事科学出版社 [Military Science Publishing House], 2013), 243.
5   Daniel Tobin, "World Class: The Logic of China's Strategy and Global Military Ambitions," in *Securing the China Dream: The PLA's Role in a Time of Reform and Change*, eds. Roy Kamphausen, David Lai and Tiffany Ma (Seattle: The National Bureau of Asian Research, 2020): 30–31.
6   Marc Lanteigne, "Beijing's Journey to the North," *The Arctic Journal*, October 8, 2015, https://web.archive.org/web/20151009165517/http://arcticjournal.com/opinion/1874/beijings-journey-north.
7   Jimu Shutuo and Qiao Heng, "5艘解放军舰艇造访白令海峡 向美国发出信息" [5 PLA Ships Visited the Bering Strait Sending a Message to the US], 环球时报 [*Global Times*], September 6, 2015, https://mil.huanqiu.com/article/9CaKrnJP8NP.
8   Shannon Tiezzi, "China's Navy Makes First-Ever Tour of Europe's Arctic States," *The Diplomat*, October 2, 2015, http://thediplomat.com/2015/10/chinas-navy-makes-first-ever-tour-of-europes-arctic-states/.
9   Andrew Higgins, "China and Russia Hold First Joint Naval Drill in the Baltic Sea," *The New York Times*, July 25, 2017, www.nytimes.com/2017/07/25/world/europe/china-russia-baltic-navy-exercises.html.
10 Yen Nee Lee, "Chinese Spy Ship Reportedly Lurking Off Coast of Alaska, Watching Anti-Missile Test," *CNBC*, July 14, 2017, www.cnbc.com/2017/07/14/china-spy-ship-reportedly-off-coast-of-alaska-watching-thaad-tests.html.
11 "中国海军174舰艇编队访问芬兰" [Chinese Navy's 174th Group Formation Visited Finland], 中国新闻网 [*China News*], August 1, 2017, www.chinanews.com/mil/2017/08-01/8293309.shtml.
12 "中俄海上联合2017军事演习圆满结束" [The Sino-Russian 2017 Joint Sea Military Exercise Came to a Successful Conclusion], *Xinhua*, September 25, 2017, www.xinhuanet.com/world/2017-09/25/c_1121721576.htm.
13 "中国海军第二十六批护航编队访问丹麦" [The Chinese Navy's 26th Escort Fleet Visited Denmark], 中国新闻网 [*China News*], September 26, 2017, www.chinanews.com/mil/2017/09-26/8340575.shtml.
14 Mark Thiessen, "Coast Guard Spots Chinese, Russian Naval Ships Off Alaska Island," *The Associated Press*, September 27, 2022, https://apnews.com/article/russia-ukraine-china-alaska-honolulu-coast-guard-54638cccc30d5a0f8879022f493a6302.
15 Jimu Shutuo and Qiao Heng, "5艘解放军舰艇造访白令海峡 向美国发出信息" [5 PLA Ships Visited the Bering Strait Sending a Message to the US].
16 Shou, 战略学 [*The Science of Military Strategy*], 246.
17 Ibid., 246–247.
18 Rasmus Gjedssø Bertelsen, "Arctic Security in International Security," in *Routledge Handbook of Arctic Security*, ed. Gunhild Hoogensen Gjørv, Marc Lanteigne and Horatio Sam-Aggrey (London: Routledge, 2020): 57–68.
19 For examples, see Xiao Tianliang et al., *The Science of Military Strategy 2020*, trans. China Aerospace Studies Institute (Montgomery, AL: China Aerospace Studies Institute, 2022), 163, and Yang Zhirong, "北极航道全年开通后世界地缘战略格局

的变化研究" [Changes in World Geo-strategic Situation and Countermeasures after the All-year-open of Arctic Channel], 国防科技 *[National Defence Science & Technology]* 36, no. 2 (2015): 8.

20 Lu Junyuan and Zhang Xia, 中国北极权益与政策研究 *[China's Arctic Interests and Policy]* (Beijing: 时事出版社 [Current Affairs Press], 2016), 27.

21 Xiao et al., *The Science of Military Strategy 2020*, 163.

22 Zhang Tong, "Chinese Scientists Simulate Hypersonic Flight to US after Devising Beidou Satellite-Switching System," *South China Morning Post*, July 18, 2022, www.scmp.com/news/china/science/article/3185709/chinese-scientists-simulate-hypersonic-flight-us-after-devising.

23 Zuo Pengfei, 极地战略问题研究 *[A Study on Polar Strategy]* (Beijing: 时事出版社 [Current Affairs Press], 2018), 13.

24 Christopher Weidacher Hsiung, "China's Technology Cooperation with Russia: Geopolitics, Economics, and Regime Security," *The Chinese Journal of International Politics* (2021): 457.

25 Hsiung, "China's Technology Cooperation with Russia: Geopolitics, Economics, and Regime Security," 457.

26 Lu and Zhang, 中国北极权益与政策研究 *[China's Arctic Interests and Policy]*, 28.

27 Qu Tanzhou, et al. 北极问题研究 *[Research on Arctic Issues]* (Beijing: 海洋出版社 [Ocean Press], 2011): 292.

28 Qu, et al. 北极问题研究 *[Research on Arctic Issues]*, 293.

29 Lu and Zhang, 中国北极权益与政策研究 *[China's Arctic Interests and Policy]*, 28.

30 For example, see Xiao et al., *The Science of Military Strategy 2020*, 163; Lu and Zhang, 中国北极权益与政策研究 *[China's Arctic Interests and Policy]*, 28–29; Qu, et al. 北极问题研究 *[Research on Arctic Issues]*, 293–294; Yang, "北极航道全年开通后世界地缘战略格局的变化研究" [Changes in World Geo-strategic Situation and Countermeasures after the All-year-open of Arctic Channel], 8; Pan Zhengxiang and Zheng Lu, "北极地区的战略价值与中国国家利益研究" [Research on the Strategic Value of the Arctic Region and China's National Interests], 江淮论坛 *[Jianghuai Forum]* 2 (2013): 121.

31 Xiao et al., *The Science of Military Strategy 2020*, 163.

32 Ibid.

33 Yang, "北极航道全年开通后世界地缘战略格局的变化研究" [Changes in World Geo-strategic Situation and Countermeasures after the All-year-open of Arctic Channel], 8.

34 For an overview of those statements see Ryan D. Martinson, "The Role of the Arctic in Chinese Naval Strategy," *China Brief* 19, no. 22 (2019), https://jamestown.org/program/the-role-of-the-arctic-in-chinese-naval-strategy/.

35 Anne-Marie Brady, *China as a Polar Great Power* (Cambridge: Cambridge University Press, 2017), 79–87.

36 Zuo, 极地战略问题研究 *[A Study on Polar Strategy]*, 14–15.

37 For an overview of these options see Adam Lajeunesse and Timothy Choi, "Here There Be Dragons? Chinese Submarine Options in the Arctic," *Journal of Strategic Studies* (2021): 7–8, https://doi.org/10.1080/01402390.2021.1940147.

38 Lajeunesse and Choi, "Here There Be Dragons? Chinese Submarine Options in the Arctic," 16.

39 Ibid., 9–16.

40 Zhang Yi, Xu Jun and Li Longyi, "把人民军队建设成为世界一流军队" [We Will Build the PLA into a World-Class Military], 人民网 *[People's Daily Online]*, September 24, 2022, http://politics.people.com.cn/n1/2022/0924/c1001-32533073.html.

41 M. Taylor Fravel, "China's 'World-Class Military' Ambitions: Origins and Implications," *The Washington Quarterly* 43, no. 1 (2020): 85–99.

42 Ibid., 91.

43 Andrew Chuter, "New Strategy Sharpens UK Military Focus on the Arctic," *Defence News*, March 29, 2022, www.defensenews.com/global/europe/2022/03/29/new-strategy-sharpens-uk-military-focus-on-the-arctic/.

44 James Holmes, "When China Rules the Sea," *Foreign Policy*, September 23, 2015, http://foreignpolicy.com/2015/09/23/when-china-rules-the-sea-navy-xi-jinping-visit/.

45 He Jian and Liu Lei, "总体国家安全观视角中的北极通道安全" [The Security of the Arctic Passage as Seen in Light of General National Security], 国家安全研究 [*Journal of International Security Studies*], no. 6 (2015): 149.

46 Zhang Hui, "China, Russia Begin Naval Drills in Sea of Japan," *Global Times*, September 18, 2017, https://web.archive.org/web/20170923142453/www.globaltimes.cn/content/1067036.shtml.

47 "New Icebreaker Joins PLA Navy in Liaoning," *China Military Online*, January 5, 2016, https://web.archive.org/web/20160216124455/english.chinamil.com.cn/news-channels/china-military-news/2016-01/05/content_6844720.htm.

48 Chen Xi, "北极哨兵：挺立在风雪中的北极杨" [The Arctic Guard: Standing Tall in the Snowstorm], *Xinhua*, February 10, 2018, www.mod.gov.cn/education/2018-02/10/content_4804596.htm.

49 Hilde-Gunn Bye, "China Will Test Winter Skills During Military Competition in Russia," *High North News*, April 9, 2021, www.highnorthnews.com/en/china-will-test-winter-skills-during-military-competition-russia.

50 Kristin Huang, "China Sends Soldiers to Russia for Snowy Mountain Challenge," *South China Morning Post*, April 1, 2021, www.scmp.com/news/china/military/article/3127770/china-sends-soldiers-russia-snowy-mountain-challenge.

51 Ryan Dean and P. Whitney Lackenbauer, "China's Arctic Gambit? Contemplating Possible Strategies," *NAADSN Policy Brief* (April 2020): 6, www.naadsn.ca/wp-content/uploads/2020/04/20-apr-23-China-Arctic-Gambit-RD-PWL-1.pdf.

52 Ibid., 6–7.

53 Nadège Rolland, "A New Great Game? Situating Africa in China's Strategic Thinking," *NBR Special Report* 91 (June 2021): 17.

54 Ibid., 17.

55 Lajeunesse and Choi, "Here There Be Dragons? Chinese Submarine Options in the Arctic," 16–17.

56 Elizabeth Wishnick, "The China-Russia 'No Limits' Partnership Is Still Going Strong, with Regime Security as Top Priority," *South China Morning Post*, September 29, 2022, www.scmp.com/comment/opinion/article/3193703/china-russia-no-limits-partnership-still-going-strong-regime.

57 Jo Inge Bekkevold, "Imperialist Master, Comrade in Arms, Foe, Partner, and Now Ally? China's Changing Views of Russia," in *Russia-China Relations: Emerging Alliance or Eternal Rivals?* eds. Sarah Kirchberger, Svenja Sinjen and Nils Wörmer (Cham: Springer, 2022), 41–57.

58 Lajeunesse and Choi, "Here There Be Dragons? Chinese Submarine Options in the Arctic," 17.

59 Ibid., 18.

60 Torbjørn Pedersen, "Polar Research and the Secrets of the Arctic," *Arctic Review on Law and Politics* 10 (2019): 110–111.

61 Yin Li et al., 极地水声信号处理研究 [Research on Underwater Signal Processing in Arctic Region], 中国科学院院刊 [*Bulletin of Chinese Academy of Sciences*] 34, no. 3 (2019): 307.

62 Pedersen, "Polar Research and the Secrets of the Arctic," 114.

63 Ibid., 104.

64 Xiao et al., *The Science of Military Strategy 2020*, 166.

65 In this example, Deng specifically mentions the Global Commons which in his analysis also include the Arctic region, see Deng Beixi, 北极安全研究 [*Arctic Security Studies*] (Beijing: 海洋出版社 [Ocean Press], 2020), 201.

66 Xiao Yang, "地缘科技学与国家安全：中国北极科考的战略深意 [Science and Technology and National Security: The Strategic Significance of China's Arctic

Scientific Expedition]," 国际安全研究 [*Journal of International Security Studies*] 33, no. 6 (2015): 107–108.

67 Alan K. Henrikson, "The Arctic Peace Projection: From Cold War Fronts to Cooperative Fora," in *Routledge Handbook of Arctic Security*, eds. Gunhild Hoogensen Gjørv, Marc Lanteigne and Horatio Sam-Aggrey (London: Routledge, 2020), 15.

68 Ronald E. Doel, "Defending the North American Continent: Why the Physical Environmental Sciences Mattered in Cold War Greenland," in *Exploring Greenland: Cold War Science and Technology on Ice*, eds. Ronald E. Doel, Kristine C. Happer and Matthias Heymann (New York: Palgrave Macmillan, 2016), 27.

69 Ibid.

70 Henrik Knudsen, "Battling the Aurora Borealis: The Transnational Coproduction of Ionospheric Research in Early Cold War Greenland," in *Exploring Greenland: Cold War Science and Technology on Ice*, eds. Ronald E. Doel, Kristine C. Happer and Matthias Heymann (New York: Palgrave Macmillan, 2016), 143–166.

71 Janet Martin-Nielsen, "Security and the Nation: Glaciology in Early Cold War Greenland," in *Exploring Greenland: Cold War Science and Technology on Ice*, eds. Ronald E. Doel, Kristine C. Happer and Matthias Heymann (New York: Palgrave Macmillan, 2016), 102.

72 Doel, "Defending the North American Continent: Why the Physical Environmental Sciences Mattered in Cold War Greenland," 27. For the successes as well as blunders of the Soviet Arctic engagement during the 1930s see John McCannon, *Red Arctic: Polar Exploration and the Myth of the North in the Soviet Union, 1932–1939* (New York and Oxford: Oxford University Press, 1998).

73 Peter Kikkert and P. Whitney Lackenbauer, "The Militarization of the Arctic to 1990," in *The Palgrave Handbook of Arctic Policy and Politics*, eds. Ken S. Coates and Carin Holroyd (Cham, Switzerland: Palgrave Macmillan, 2020), 495–496.

74 Doel, "Defending the North American Continent: Why the Physical Environmental Sciences Mattered in Cold War Greenland," 27.

75 Pedersen, "Polar Research and the Secrets of the Arctic," 128.

76 "China's Military Strategy (2015)," *Xinhua*, May 27, 2015, http://english.www.gov.cn/archive/white_paper/2015/05/27/content_281475115610833.htm.

77 Stephen Chen, "US Spy Planes Kept Eye on Chinese Scientists During Research Expedition Near Guam," *South China Morning Post*, October 5, 2017, www.scmp.com/news/china/diplomacy-defence/article/2113883/us-spy-planes-kept-eye-chinese-scientists-during.

78 James Kynge et al., "How China Rules the Waves," *Financial Times*, January 12, 2017, https://ig.ft.com/sites/china-ports/.

79 Isaac B. Kardon and Wendy Leutert, "Pier Competitor: China's Power Position in Global Ports," *International Security* 46, no. 4 (2022): 9–47.

80 Xiao et al., *The Science of Military Strategy 2020*, 165.

81 Deng, 北极安全研究 [*Arctic Security Studies*], 236.

82 Ryan D. Martinson and Peter A. Dutton, "China Maritime Report No. 3: China's Distant-Ocean Survey Activities: Implications for US National Security," *CMSI China Maritime Reports* 3, (2018): 11.

83 Ibid.

84 Brady, *China as a Polar Great Power*, 102–107; Rush Doshi, Alexis Dale-Huang and Gaoqi Zhang, *Northern Expedition: China's Arctic Activities and Ambitions* (Washington DC: The Brookings Institution, 2021), 29–35.

85 For example, see Wei Zexun, ed., 中国第九次北极科学考察报告 [*The Report of 2018 Chinese Arctic Research Expedition*] (Beijing: 海洋出版社 [Ocean Press], 2019).

86 "Fields of Science," Arctic Observatory, n.d., https://karholl.is/en/science.

87 Qu, et al. 北极问题研究 [*Research on Arctic Issues*], 296.

88 Ibid.

89 Martinson and Dutton, "China Maritime Report No. 3: China's Distant-Ocean Survey Activities: Implications for US National Security," 12.
90 Unfortunately, at the time of the writing of this chapter, the website of the office was offline. The author does however have file snapshots from the webpage describing the main responsibilities of the office.
91 Martinson and Dutton, "China Maritime Report No. 3: China's Distant-Ocean Survey Activities: Implications for US National Security," 12–13.
92 "概况" [Overview], 自然资源部第三海洋研究所 [Third Institute of Oceanography of the MNR], n.d., www.tio.org.cn/OWUP/html/gk.html.
93 Bryan J.R. Millard and P. Whitney Lackenbauer, *Trojan Dragons? Normalizing China's Presence in the Arctic* (Calgary: Canadian Global Affairs Institute, 2021), 19.
94 Ibid.
95 Rebecca Pincus, "Towards a New Arctic," *The RUSI Journal 165*, no. 3 (2020): 54.
96 "中国第十次北极考察队返回青岛" [China's 10th Arctic Scientific Expedition Returned to Qingdao], 自然资源部第一海洋研究所 [The First Institute of Oceanography], September 30, 2019, www.mnr.gov.cn/dt/ywbb/201909/t20190930_2469709.html.
97 Wang Jing, "向阳红01船起航执行中国第十次北极考察任务" [Xiang Yang Hong 1 Set Sail for China's 10th Arctic Expedition], 中国海洋报 [*China Ocean News*], August 12, 2019, https://hyda.nmdis.org.cn/c/2019-08-12/68236.shtml.
98 Martinson and Dutton, "China Maritime Report No. 3: China's Distant-Ocean Survey Activities: Implications for US National Security," 6–7.
99 H.I. Sutton, "Chinese Ships Seen Mapping Strategic Seabed In Indian Ocean," Naval News, January 22, 2021, www.navalnews.com/naval-news/2021/01/how-china-is-mapping-the-seabed-of-the-indian-ocean/.
100 Ibid.
101 Brady, *China as a Polar Great Power,* 107.
102 "系统介绍" [System Introduction], 北斗卫星导航系统 [Beidou Navigation Satellite System], n.d., www.beidou.gov.cn/xt/xtjs/.
103 Andrew S. Erickson and Conor M. Kennedy, "China's Maritime Militia," in *Becoming a Great Maritime Power: A Chinese Dream,* ed. Michael McDevitt (Arlington, CAN Analysis and Solutions, 2016), 72–73.
104 Samuel Strickland, "How China's Military Plugs Into the Global Space Sector," *The Strategist*, October 26, 2022, www.aspistrategist.org.au/how-chinas-military-plugs-into-the-global-space-sector/.
105 Yan Ronghua, "国家测绘地理信息局北极黄河站北斗卫星导航定位基准站开通运行" [State Bureau of Surveying and Mapping's Yellow River Station Beidou Satellite Navigation and Positioning Reference Station Began Operations], 国家测绘地理信息局 [*State Bureau of Surveying and Mapping*], September 1, 2016, www.mnr.gov.cn/dt/ch/201609/t20160901_2345839.html.
106 Stephen Chen, "China Launches its First Fully Owned Overseas Satellite Ground Station Near North Pole," *South China Morning Post*, December 16, 2016, www.scmp.com/news/china/policies-politics/article/2055224/china-launches-its-first-fully-owned-overseas-satellite.
107 Ding Jia, "中国遥感卫星地面站北极站投入试运行" [China's Arctic Remote Sensing Satellite Ground Station Was Put into Trial Operations], 科学网 [*Science Net*], December 15, 2016, http://news.sciencenet.cn/htmlnews/2016/12/363617.shtm.
108 Mark Stokes and Ian Easton, "China's Evolving Space Capabilities: Implications for US Interests," in *Routledge Handbook of Chinese Security*, ed. Lowell Dittmer and Yu Maochun (London and New York: Routledge, 2015), 334.
109 Keegan Elmer, "Swedish Defence Agency Warns Satellite Station Could Be Serving Chinese Military," *South China Morning Post*, January 14, 2019, www.scmp.com/news/china/diplomacy/article/2182026/swedish-defence-agency-warns-satellite-station-could-be-serving.

110 Jonathan Barrett and Johan Ahlander, "Exclusive: Swedish Space Company Halts New Business Helping China Operate Satellites," *Reuters*, September 21, 2020, www.reuters.com/article/uk-china-space-australia-exclusive-idUKKCN26C20A.

111 Cui Xuejin, "北师大在格陵兰启动北极第二个卫星地面站建设" [Beijing Normal University Initiated Construction of the Second Arctic Satellite Ground Receiving Station in Greenland], 科学网 [*Science Net*], June 4, 2017, http://news.sciencenet.cn/htmlnews/2017/6/378159.shtm.

112 "中俄卫星导航重大战略合作项目委员会第九次会议成功举行" [The Ninth Meeting of the China-Russia Satellite Navigation Major Strategic Cooperation Project Committee Was Successfully Held], 北斗卫星导航系统 [*Beidou Navigation Satellite System*], September 27, 2022, www.beidou.gov.cn/yw/xwzx/202209/t20220929_24623.html.

113 Shi Changxue, "中国首赴南极考察：海军借此培养远航指挥人才" [China's First Antarctic Expedition: The Navy Uses It for Training Officers in Far Ocean Sailing], 中国海洋报 [*China Ocean News*], April 24, 2013, www.chinanews.com/mil/2013/04-24/4760323.shtml.

114 Hu Baoliang, "第一个在南极上空飞行的中国飞行员" [The First Chinese Pilot to Fly in Antarctica], 中国海军网 [*Chinese Navy Online*], August 10, 2017, https://web.archive.org/web/20170818123125/http://navy.81.cn/content/2017-08/10/content_7713845.htm.

115 "中国第1次南极科学考察" [China's First Antarctic Scientific Expedition], 极地之门 [*Gate to the Poles*], n.d., https://web.archive.org/web/20211205074110/http://polar.org.cn/expeditionDetail/?id=781.

116 Shi, "中国首赴南极考察：海军借此培养远航指挥人才" [China's First Antarctic Expedition: The Navy Uses It for Training Officers in Far Ocean Sailing].

117 Xiao et al., *The Science of Military Strategy 2020*, 165.

118 Ibid.

119 Yin et al., 极地水声信号处理研究 [Research on Underwater Signal Processing in Arctic Region], 309.

120 Song Yunxia, "极地军事活动概览" [Overview of Polar Military Activities], in 极地法律问题 [*Legal Issues of the Polar Regions*], ed. Jia Yu (Beijing: 社会科学文献出版社 [Social Sciences Academic Press], 2014), 176–177.

121 Xiao et al., *The Science of Military Strategy 2020*, 166.

122 Ibid.

123 Fan Chuangui, "经济安全缘何被视为国家安全基础" [Why Economic Security Is Considered the Foundation of National Security], 法制日报 [*Legal Daily*], April 22, 2014, http://theory.people.com.cn/n/2014/0422/c40531-24927787.html.

124 "总体国家安全观的16种安全" [16 Kinds of Security of the Comprehensive National Security Outlook].

125 For example, see: "中央经济工作会议在北京举行" [Central Economic Work Conference Held in Beijing], 人民日报 [*People's Daily*], December 11, 2021, http://cpc.people.com.cn/n1/2021/1211/c64094-32305295.html.

126 Jeff Barron, "China Surpassed the United States as the World's Largest Crude Oil Importer in 2017," *US Energy Information Administration*, December 31, 2018, www.eia.gov/todayinenergy/detail.php?id=37821.

127 Victoria Zaretskaya and Faouzi Aloulou, "As of 2021, China Imports More Liquefied Natural Gas Than Any Other Country," *US Energy Information Administration*, May 2, 2022, www.eia.gov/todayinenergy/detail.php?id=52258.

128 "Xi Focus: Walk the Talk: Xi Leads China in Fight for Carbon-Neutral Future," *Xinhua*, March 16, 2021, www.xinhuanet.com/english/2021-03/16/c_139814792.htm.

129 "关于印发《"十四五"能源领域科技创新规划》的通知" [Notice Regarding the Distribution of the 14th Five-Year Plan for Scientific and Technological Innovation in the Field of Energy], 国家能源局 [*National Energy Administration*], November 29, 2021, www.gov.cn/zhengce/zhengceku/2022-04/03/content_5683361.htm.

130 "Sinopec Releases China Energy Outlook 2060, Anchoring New Path of Energy Transformation Development," *Sinopec*, December 29, 2022, www.sinopecgroup.com/group/en/Sinopecnews/20221229/news_20221229_577140178923.shtml.

131 Muyu Xu and Chen Aizhu, "China's Sinopec Plans Its Biggest Capital Expenditure in History," *Reuters*, March 28, 2022, www.reuters.com/business/energy/chinas-sinopec-plans-its-biggest-capital-expenditure-history-2022-03-27/.

132 "China – Executive Summary," *US Energy Information Administration*, August 8, 2022, www.eia.gov/international/analysis/country/CHN.

133 Lin Boqiang, "促进中国油气进口多元化非常必要" [It Is Very Necessary to Advance the Diversification of China's Oil and Gas Imports], 21世纪经济报道 [*21st Century Business Herald*], February 16, 2022, https://m.21jingji.com/article/20220216/4da6454e24383f9eb9b3046410d13781.html.

134 Sun Kai and Wu Hao, "北极安全治理中的角色定位与策略选择" [China's Role and Strategic Choice in Arctic Security Governance], in 北极地区发展报告2017 [*Report on Arctic Region Development 2017*], ed. Liu Huirong (Beijing: 社会科学文献出版社 [Social Sciences Academic Press], 2017), 55–56.

135 Zhang Mengxiao and Ni Sijie, "开发北极：中国不能落下" [Developing the Arctic: China Cannot Drop Out], 中国科学报 [*China Science Daily*], September 23, 2014, https://news.sciencenet.cn/sbhtmlnews/2014/9/292227.shtm.

136 "中国极地考察十二五发展规划" [The Chinese 12th Five-Year Development Plan for Polar Expeditions], 国家海洋局 [*The State Oceanic Administration*], August 27, 2011, www.soa.gov.cn/zwgk/hygb/gjhyjgb/2011_2/201508/t20150827_39802.html. The link has been discontinued and was not saved on archiving servers such as the Wayback Machine. The author, however, has a physical copy of the development plan on file.

137 "Full Text: China's Arctic Policy," *Xinhua*.

138 Lu and Zhang, 中国北极权益与政策研究 [*China's Arctic Interests and Policy*], 30–31.

139 Wang Jiefeng, "气候变暖升高北极价值" [Climate Warming Increases the Value of the Arctic], 瞭望 [*Outlook*] no. 26 (2022), http://lw.news.cn/2022-06/27/c_1310633782.htm.

140 "Full Text: China's Arctic Policy," *Xinhua*.

141 Gao Tianming et al., "Can Russian Arctic Regions Benefit from Collaborating with North-Eastern China? Current Challenges to the Low-Carbon Agenda," in *Arctic Yearbook 2022 – The Russian Arctic: Economic, Politics & Peoples*, eds. Lassi Heininen, Heather Exner-Pirot and Justin Barnes (Akureyri, Iceland: Arctic Portal, 2022), 1–28, https://arcticyearbook.com/arctic-yearbook/2022/2022-scholarly-papers/426-can-russian-arctic-regions-benefit-from-collaborating-with-northeastern-china-current-challenges-to-the-low-carbon-agenda.

142 "极寒北极地位升温 矿产资源开发号角已吹响" [The Extremely Cold Arctic Is Warming Up, the Trumpet for Mineral Resource Development Has Been Sounded], 科技日报 [*Science and Technology Daily*], January 22, 2019, www.xinhuanet.com/politics/2019-01/22/c_1124022713.htm.

143 Liu Qingping, Zhao Yuanyi and Liu Chunhua, "格陵兰岛稀土矿资源潜力及对中国的可利用性评价" [Potential of Rare Earth Resources in Greenland and Evaluation of Its Availability to China], 地质通报 [*Geological Bulletin of China*] 38, no. 8 (2019): 1386–1395.

144 The PRC currently controls around 90% of these materials and dominates the highly specialise metallurgy and the supply chain, see: Sophia Kalantzakos, *China and the Geopolitics of Rare Earths* (New York: Oxford University Press, 2018).

145 Stacy Meichtry and Drew Hinshaw, "China's Greenland Ambitions Run into Local Politics, US Influence," *The Wall Street Journal*, April 8, 2021, www.wsj.com/articles/chinas-rare-earths-quest-upends-greenlands-government-11617807839.

146 For the "Arctic scramble" narrative see: Scott G. Borgerson, "Arctic Meltdown: The Economic and Security Implications of Global Warming," *Foreign Affairs* 87, no. 2 (2008): 63–77.

147 Jia Lingxiao, "北极地区油气资源勘探开发现状" [Current Situation of Oil and Gas Exploration and Development in the Arctic Region], 中国矿业报 [*China Mining News*], July 14, 2017, www.mnr.gov.cn/dt/kc/201707/t20170714_2322077.html.

148 Vita Spivak and Alexander Gabuev, "The Ice Age: Russia and China's Energy Cooperation in the Arctic," *Carnegie Endowment for International Peace*, December 31, 2021, https://carnegiemoscow.org/commentary/86100.

149 "亚马尔项目助力中俄能源合作稳步前行" [The Yamal Project Helps the Sino-Russian Energy Cooperation Forge Steadily Ahead], *Xinhua*, December 12, 2017, http://news.xinhuanet.com/world/2017-12/12/c_129763727.htm.

150 Spivak and Gabuev, "The Ice Age: Russia and China's Energy Cooperation in the Arctic."

151 Chen Aizhu, Julie Zhu and Muyu Xu, "Exclusive China's Sinopec Pauses Russia Projects, Beijing Wary of Sanctions-Sources," *Reuters*, March 28, 2022, www.reuters.com/business/energy/exclusive-chinas-sinopec-pauses-russia-projects-beijing-wary-sanctions-sources-2022-03-25/.

152 Malte Humpert, "China Receives Late-Season LNG Deliveries from Russian Arctic Capping Off Record-Breaking Year," *High North News*, January 2, 2023, www.highnorthnews.com/en/china-receives-late-season-lng-deliveries-russian-arctic-capping-record-breaking-year.

153 Duan Zhaofang, "健全多元化海外供应体系 保障我国天然气稳定供应" [Strengthen a Diversified Overseas Supply System to Ensure the Stable Supply of Natural Gas in My Country], 中国石油新闻中心 [*CNPC News Centre*], November 19, 2019, http://news.cnpc.com.cn/system/2019/11/19/001752319.shtml.

154 Zhang Xiaodong, "冰上丝路见证中俄合作新成果" [The Polar Silk Road Witnessed New Achievements in Sino-Russia Cooperation], 人民日报 [*People's Daily*], December 11, 2017, https://web.archive.org/web/20230212170312/http://paper.people.com.cn/rmrb/html/2017-12/11/nw.D110000renmrb_20171211_1-21.htm.

155 For example, see: Yang Jian, "北极航运与中国北极政策定位" [Arctic Shipping and China's Arctic Policy Orientation], 国际观察 [*International Review*], no. 1 (2014): 123–137, Xiao Yang, "北冰洋航线开发:中国的机遇与挑战" [The Development of the Arctic Sea Routes: Opportunities and Challenges for China], 现代国际关系 [*Contemporary International Relations*], no. 6 (2011): 52–57; and Liu Xingpeng, "北极航线对我国海运强国的战略价值" [The Strategic Value of Arctic Shipping Routes for China as a Maritime Transport Power], 中国港口 [*China Ports*], no. 7 (2017), www.sohu.com/a/163169869_784079.

156 For example, see: Xu Yang and Zou Mingzhong, "北极航道有望成为繁荣中欧贸易新的黄金水道" [The Arctic Sea Route Is Expected to Become a New Golden Waterway for the Prosperous Trade Between China and Europe], *Xinhua*, October 10, 2015, www.xinhuanet.com/world/2015-10/24/c_128354115.htm.

157 Qiang Zhang, Zheng Wan and Shanshan Fu, "Toward Sustainable Arctic Shipping: Perspectives from China," *Sustainability* 12 (2020): 8.

158 "Keynote Speech by Vice Foreign Minister Zhang Ming at the China Country Session of the Third Arctic Circle Assembly," *Ministry of Foreign Affairs of the PRC*, October 17, 2015, www.fmprc.gov.cn/eng/wjdt_665385/zyjh_665391/201510/t20151017_678393.html.

159 Marc Lanteigne, "Who Benefits From China's Belt and Road in the Arctic?" *The Diplomat*, September 12, 2017, https://thediplomat.com/2017/09/who-benefits-from-chinas-belt-and-road-in-the-arctic/.

160 "Full Text: China's Arctic Policy," *Xinhua*.

161 Sun Kai, "从愿景到行动：推进冰上丝绸之路建设正当其时" [From Vision to Action: The Right Time for Advancing the Polar Silk Road], 中国社会科学网 [*Chinese Social Sciences Net*], February 8, 2018, https://web.archive.org/web/20210717231948/www.cssn.cn/zzx/gjzzx_zzx/201802/t20180208_3845186.shtml.

162 For a discussion on the potential development of the Transpolar Sea Route, see Mia M. Bennett et al., "The Opening of the Transpolar Sea Route: Logistical, Geopolitical, Environmental, and Socioeconomic Impacts," *Marine Policy* 121 (2020): 1–15.

163 Chen Jinlei et al., "Projected Changes in Sea Ice and the Navigability of the Arctic Passages Under Global Warming of 2 ℃ and 3 ℃," *Anthropocene* 40 (2022), https://doi.org/10.1016/j.ancene.2022.100349.

164 Zhang Yiqian, "征服北极黄金水道 探秘中企船队开拓冰上丝绸之路的九九八十一难" [Conquering the Arctic Golden Waterway, Exploring the Difficulties of the Chinese Enterprise Fleet in Opening the Polar Silk Road], 环球时报 [*Global Times*], February 12, 2018, https://world.huanqiu.com/article/9CaKrnK6G7k.

165 Lawson W. Brigham, "Arctic Shipping Routes: Russia's Challenges and Uncertainties," *The Barents Observer*, August 12, 2022, https://thebarentsobserver.com/en/opinions/2022/08/arctic-shipping-routes-russias-challenges-and-uncertainties.

166 Björn Gunnarsson and Arild Moe, "Ten Years of International Shipping on the Northern Sea Route: Trends and Challenges," *Arctic Review on Law and Politics* 12 (2021): 22.

167 Xu Ren, ed., 中国第八次北极科学考察报告 [*The Report of 2017 Chinese National Arctic Research Expedition*] (Beijing: 海洋出版社 [Ocean Press], 2019), 330.

168 Liang Xi et al., "Evaluation of ArcIOPS Sea Ice Forecasting Products During the Ninth Chinare-Arctic in Summer 2018," *Advances in Polar Science* 31, no. 1 (2020): 14–25.

169 Zhang Peng and Pang Haoyan, "全球首部中文版《北极航行指南（西北航道）》将发布" [The World's First Chinese Navigation Guidelines for the Arctic (Northwest Passage) Will Be Released], 中国新闻网 [*China News*], December 3, 2015, https://world.huanqiu.com/article/9CaKrnJRXNc.

170 Ma Yanling and Liu Jiayu, "北极航海地图集正式发行" [Atlas for Arctic Navigation Was Officially Issued], 中国新闻网 [*China News*], October 29, 2015, www.chinanews.com/gn/2015/10-29/7596347.shtml.

171 Zhou Runjian, "我国编制完成《北极东北航道通信指南》" [China Compiled Arctic Northeast Passage Communication Guidelines], *Xinhua*, December 23, 2017, https://web.archive.org/web/20171225213109/www.xinhuanet.com/fortune/2017-12/23/c_1122157172.htm.

172 "新版《北极航行指南（东北航道）》正式出版发行" [A New Version of the Navigation Guidelines for the Arctic (Northeast Passage) Were Published], 中国水运报 [*China Water Transport*], July 28, 2022, www.msa.gov.cn/html/xxgk/hsyw/20220728/46BC9AE6-0371-4121-BBC4-72598FCD4343.html.

173 Yang Chupeng and Liu Weihong, "北极区域地质与油气资源专著出版发行" [The Monograph on Arctic Geology and Oil and Gas Resources Was Published], 广州海洋地质调查局 [*Guangzhou Marine Geological Survey*], December 23, 2014, www.cgs.gov.cn/xwl/cgkx/201603/t20160309_297409.html.

174 "国家安全之科技安全" [National Security's Technological Security], 湖口县人民政府 [*Hukou County People's Government*], April 14, 2022, https://web.archive.org/web/20221222093124/www.hukou.gov.cn/kjj/gzdt/202204/t20220414_5456129.html.

175 Ibid.

176 Spivak and Gabuev, "The Ice Age: Russia and China's Energy Cooperation in the Arctic."

177 "亚马尔项目助力中俄能源合作稳步前行" [The Yamal Project Helps the Sino-Russian Energy Cooperation Forge Steadily Ahead], *Xinhua*, December 12, 2017, www.xinhuanet.com//world/2017-12/12/c_129763727.htm.

178 Luan Hai, "专访：科技工程助力北极油气开发 – 访中石油俄罗斯公司总经理蒋奇" [Interview: General Manager of CNPC Russia Jian Qi – Science and Technology Projects Help Arctic Oil and Gas Development], *Xinhua*, June 13, 2017, https://web.archive.org/web/20190605211859/www.xinhuanet.com/fortune/2017-06/13/c_1121134481.htm.

179 "China Signs $1.6 bln Engineering Deal for Siberian LNG Project," *Reuters*, July 10, 2014, www.reuters.com/article/china-gas-yamal-idUSL4N0PL20C20140710.

180 Spivak and Gabuev, "The Ice Age: Russia and China's Energy Cooperation in the Arctic."

181 Wang Mingping, "抗风12级抗冻-60℃，四川造极地钻机北极圈大显身手" [Wind Resistance Level 12, Freeze Resistance −60°C, Made in Sichuan Polar Drilling Rig Shows Its Skills in the Arctic Circle], 红星新闻 [*Hong Xing Xin Wen*], March 7, 2019, https://ishare.ifeng.com/c/s/7kqXXozm725.

182 "中国化学工程签约俄罗斯帕亚哈油气田项目合作协议" [China Chemical Engineering Signed a Cooperation Agreement on the Payaha Oil and Gas Field Project in Russia], 中国化学工程集团 [*China National Chemical Engineering*], June 8, 2019, www.sasac.gov.cn/n2588025/n2588124/c11445784/content.html.

183 Liu Zhen, "China to Develop New Heavy Icebreaker for Polar Silk Road," *South China Morning Post*, November 13, 2021, www.scmp.com/news/china/diplomacy/article/3155860/china-develop-new-heavy-icebreaker-polar-silk-road.

184 "交通运输部关于部救助打捞局开展重型破冰救助船研究等交通强国建设试点工作的意见" [Suggestions of the Ministry of Transport on the Rescue and Salvage Bureau's Research on Heavy-Duty Icebreaking Ship and Other Pilot Work on Building China into a Transportation Great Power], 中华人民共和国交通运输部 [*Ministry of Transport of the PRC*], October 26, 2021, https://xxgk.mot.gov.cn/2020/jigou/zh-ghs/202110/t20211026_3623048.html.

185 For an excellent analysis of the PRC's nuclear-powered icebreaker and related technology ambitions see Trym Aleksander Eiterjord, "Checking in on China's Nuclear Icebreaker," *The Diplomat*, September 5, 2019, https://thediplomat.com/2019/09/checking-in-on-chinas-nuclear-icebreaker/.

186 "中国新型极地开发破冰船提出初步方案" [Preliminary Plan for China's New Polar Icebreaker Was Proposed], 中国新闻网 [*China News*], March 17, 2014, www.chinanews.com/mil/2014/03-17/5959177.shtml.

187 "国务院关于印发"十三五"国家战略性新兴产业发展规划的通知" [The State Council Issued a Notice on the Development Plan of the Strategic Emerging Industries in the 13th Five-Year Plan], 中华人民共和国中央人民政府 [*The State Council of the PRC*], December 19, 2016, www.gov.cn/zhengce/content/2016-12/19/content_5150090.htm.

188 Eiterjord, "Checking in on China's Nuclear Icebreaker."

189 "上海交通大学与中国核工业集团有限公司签署深化战略合作协议" [Shanghai Jiaotong University and the China National Nuclear Corporation Signed an Agreement to Deepen Strategic Cooperation], 上海交通大学 [*Shanghai Jiaotong University*], August 13, 2018, https://news.sjtu.edu.cn/jdyw/20180815/81549.html.

190 Eiterjord, "Checking in on China's Nuclear Icebreaker." For the Chinese source see Chen Yu, "我国首艘核动力破冰船揭开面纱 – 将为海上浮动核电站动力支持铺开道路" [China's First Nuclear-Powered Icebreaker Has Been Unveiled – Paving the Way for a Floating Nuclear Power Plant], 科技日报 [*Science and Technology Daily*], June 27, 2018, https://web.archive.org/web/20210415004740/www.stdaily.com/kjrb/kjrbbm/2018-06/27/content_684661.shtml.

191 Office of the Secretary of Defence, *Annual Report to Congress: Military and Security Developments Involving the People's Republic of China 2017* (Washington, DC: Department of Defence, 2017), 69.

192 Camilla T.N. Sørensen and Christopher Weidacher Hsiung, "The Role of Technology in China's Arctic Engagement: A Means as Well as an End in Itself," in *Arctic Yearbook 2021: Defining and Mapping the Arctic: Sovereignties, Policies and Perceptions*, eds. Lassi Heininen, Heather Exner-Pirot and Justin Barnes (Akureyri, Iceland: Arctic Portal, 2021), 5, https://arcticyearbook.com/images/yearbook/2021/Scholarly-Papers/11_AY2021_Sorensen_Hsiung.pdf.

193 Xiao et al., *The Science of Military Strategy 2020*, 163.

194 Dong Yongzai, "极地安全：国家安全的新疆域" [Polar Security: A New Frontier for National Security], 光明日报 [*Guangming Daily*], April 25, 2021, https://news.gmw.cn/2021-04/25/content_34790839.htm.

195 Song, "极地军事活动概览" [Overview of Polar Military Activities], 182.

196 Ryan Martinson, "Gliders with Ears: A New Tool in China's Quest for Undersea Security," *Centre for International Maritime Security*, March 21, 2022, https://cimsec.org/gliders-with-ears-a-new-tool-in-chinas-quest-for-undersea-security/.

197 Liu Kui and Ma Xiaojing, "无人潜航器，未来水下战场的黑马" [Unmanned Underwater Vehicles: The Black Horses of the Underwater Warfare of the Future], 中国青年报 [*China Youth Daily*], September 6, 2018, http://zqb.cyol.com/html/2018-09/06/nw.D110000zgqnb_20180906_3-12.htm.

198 Hsiung, "China's Technology Cooperation with Russia: Geopolitics, Economics, and Regime Security," 460.

199 Liu Zhenglu, "8国加强北极军事部署：冰原利器征战寒荒" [8 Countries Strengthening Military Deployment in the Arctic: Ice Weapons Conquering the Cold Desert], 中国军网 [*China Military*], June 1, 2018, https://web.archive.org/web/20220330085600/www.81.cn/jwgz/2018-06/01/content_8047504.htm.

200 Yin et al., 极地水声信号处理研究 [Research on Underwater Signal Processing in Arctic Region], 309.

201 Wei Zexun, ed., 中国第九次北极科学考察报告 [*The Report of 2018 Chinese Arctic Research Expedition*] (Beijing: 海洋出版社 [Ocean Press], 2019), 45–60.

202 Liu and Ma, "无人潜航器，未来水下战场的黑马" [Unmanned Underwater Vehicles: The Black Horses of the Underwater Warfare of the Future].

203 Frank Jüris, "Chinese Security Interests in the Arctic: From Sea Lanes to Scientific Cooperation," in *Nordic-Baltic Connectivity with Asia via the Arctic: Assessing Opportunities and Risks*, eds. Bart Gaens, Frank Jüris and Kristi Raik (Tallinn, Estonia: International Centre for Defence and Security, 2021), 145.

204 "学院获批一项国家自然科学基金重大项目" [The College Was Approved a Major Project by the National Natural Science Foundation of China], 哈尔滨工程大学 [*Harbin Engineering University*], December 8, 2021, https://web.archive.org/web/20220320014419/sec.hrbeu.edu.cn/2021/1207/c257a279807/page.htm.

205 "2019首届中俄极地声学研讨会圆满举行" [The First 2019 China-Russia Symposium on Polar Acoustics Was Successfully Held], 哈尔滨工程大学 [*Harbin Engineering University*], July 22, 2019, http://uae.hrbeu.edu.cn/2019/0722/c3751a235665/page.htm.

206 "国家重点研发计划"极端环境下北极航行船舶运动和结构安全性分析与评估"项目全面启动" [The National Key R&D Program "Analysis and Evaluation of Ship Movement and Structural Safety in Arctic Navigation in Extreme Environment" Was Fully Launched], 哈尔滨工程大学 [*Harbin Engineering University*], May 29, 2018, http://sec.hrbeu.edu.cn/2018/0529/c441a190142/page.htm.

207 Sørensen and Hsiung, "The Role of Technology in China's Arctic Engagement: A Means as Well as an End in Itself," 9.

208 Ibid.

209 For the development of the modern industrial cluster in the PRC, see: Wang Zheng, "全力推动工业经济积极恢复、稳步回升（权威访谈）–访工业和信息化部党组书记、部长金壮龙" [Fully Promote an Active and Steady Recovery of the Industrial Economy (Authoritative Interview) – Interview with Party Secretary and Minister Jin Zhuanglong of the Ministry of Industry and Information Technology], 人民日报 [*People's Daily*], January 3, 2023, http://paper.people.com.cn/rmrb/html/2023-01/03/nw.D110000renmrb_20230103_2-02.htm.

210 For example, see: Zhao Yue et al., "冰上丝绸之路与北极油气资源" [Polar Silk Road and Arctic Petroleum and Gas Resources], 地质力学学报 [*Journal of Geomechanics*] 27, no. 5 (2021): 888.

211 Camilla T.N. Sørensen, "China and the Arctic: Establishing Presence and Influence," in *Hybrid CoE Research Report 4: Security and Hybrid Threats in the Arctic: Challenges*

and Vulnerabilities of Securing the Transatlantic Arctic, ed. Paul Dickson and Anna-Kaisa Hiltunen (Helsinki: Hybrid CoE, 2021), 23.

212 Yang Dazhi, "政治安全是国家安全的根本" [Political Security Is the Root of National Security], 解放军报 [*PLA Daily*], April 20, 2018, www.mod.gov.cn/gfbw/jmsd/4809950.html.

213 "习近平新时代中国特色社会主义思想学习问答42" [Study Questions and Answers Regarding Xi Jinping Thought on Socialism with Chinese Characteristics in the New Era 42], 人民日报 [*People's Daily*], September 14, 2021, http://politics.people.com.cn/n1/2021/0914/c1001-32225844.html.

214 Camilla T.N. Sørensen, "Is China Becoming More Aggressive? A Neoclassical Realist Analysis," *Asian Perspectives*, no. 37 (2013): 376.

215 Erica Pandey, "China's Rust Belt," *Axios*, September 12, 2019, www.axios.com/2019/09/12/northeast-china-rust-belt.

216 Frank Tang, "China Fighting Uphill Battle to Revitalise Northeast Rust-Belt in Latest Bid to Transform Former Industrial Heartland," *South China Morning Post*, August 25, 2021, www.scmp.com/economy/china-economy/article/3146367/china-fighting-uphill-battle-revitalise-northeast-rust-belt.

217 "国务院办公厅关于调整国务院振兴东北地区等老工业基地领导小组组成人员的通知" [Circular of the General Office of the State Council on the Adjustment of the Composition of the State Council Leading Small Group to Revitalize the Old Industrial Bases of Northeast China], 中华人民共和国中央人民政府 [*The State Council of the PRC*], July 8, 2013, www.gov.cn/zwgk/2013-07/08/content_2442633.htm.

218 "Vision and Actions on Jointly Building Silk Road Economic Belt and 21st Century Maritime Silk Road," *National Development and Reform Commission*, March 28, 2015, https://web.archive.org/web/20160422233810/http://en.ndrc.gov.cn/newsrelease/201503/t20150330_669367.html.

219 Qi Haishan, Xu Yang and Li Jianping, "东北融入一带一路期待新振兴" [The Northeast Integrates into BRI Expecting Revitalization], *Xinhua*, April 1, 2015, http://politics.people.com.cn/n/2015/0401/c70731-26785531.html.

220 Martin Kossa, Marina Lomaeva and Juha Saunavaara, "East Asian Subnational Government Involvement in the Arctic: A Case for Paradiplomacy," *The Pacific Review* 34, no. 4 (2021): 664–695.

221 "辽海欧开通第二条北极航线" [Liaoning to Europe via the Ocean Opened up a Second Arctic Route], 厅政策法规处 [*Department of Policy and Regulations*], July 3, 2018, https://jtt.ln.gov.cn/jtt/fw/jtxw/414B399A24DF44BF89A4B15A17FC9AC8/index.shtml.

222 Kossa, Lomaeva and Saunavaara, "East Asian Subnational Government Involvement in the Arctic: A Case for Paradiplomacy," 679.

223 中共黑河市委宣传部 [The Propaganda Department of Heihe CCP Municipal Committee], "全力打造冰上丝绸之路黄金支点" [Spare No Effort Building the Golden Fulcrum of the Polar Silk Road], 奋斗 [*Struggle*] 2 (2018): 21–22.

224 Kossa, Lomaeva and Saunavaara, "East Asian Subnational Government Involvement in the Arctic: A Case for Paradiplomacy," 684.

225 Deborah Welch Larson, T.V. Paul and William C. Wohlforth, "Status and World Order," in *Status in World Politics*, eds. T.V. Paul, Deborah Welch Larson and William C. Wohlforth (Cambridge: Cambridge University Press, 2014), 7.

226 Deborah Welch Larson and Alexei Shevchenko, "Chinese and Russian Responses to US Primacy," *International Security* 34, no. 4 (2010): 63–95.

227 Larson, Paul and Wohlforth, "Status and World Order," 11.

228 Robert S. Ross, "China's Naval Nationalism: Sources, Prospects, and the US Response," *International Security* 34, no. 2 (2009): 63.

229 Pu Xiaoyu and Randall L. Schweller, "Status Signalling, Multiple Audiences, and China's Blue-Water Naval Ambition," in *Status in World Politics*, eds. T.V. Paul, Deborah

Welch Larson and William C. Wohlforth (Cambridge: Cambridge University Press, 2014), 145.
230 Aki Tonami, "Influencing the Imagined Polar Regions: The Politics of Japan's Arctic and Antarctic Policies," *Polar Record* 53, no. 5 (2017): 489–490.
231 Li Dahai and Zhang Yingnan, "冰上丝绸之路海洋科技创新战略研究" [Marine Science and Technology Innovation for the Polar Silk Road], 中国工程科学 [*Strategic Study of Chinese Academy of Engineering*] 21, no. 6 (2019): 66.
232 Lajeunesse and Choi, "Here There Be Dragons? Chinese Submarine Options in the Arctic," 16.

5 The PRC and Arctic Governance

Chinese Normative Discourses, Solutions, and Strategic Options

Under the leadership of General Secretary Xi, the PRC is deepening its engagement in global governance and further stepping up activism on the international stage. In a world characterised by turmoil and uncertainty, the Chinese government promotes its unique solutions, such as the Global Security Initiative and the Global Development Initiative, with the aim of reforming the current global governance system, which is still dominated by Western powers, to make it more just and equitable.[1] This activism reflects China's growing comprehensive national power, its global significance, and its willingness to shape the agendas of global governance in order to enhance its discourse power and foster a pluralistic and inclusive approach to shared responsibilities.[2] Chinese efforts to shape aspects of the current global governance system are particularly relevant when it comes to the strategic new frontiers, including outer space, cyberspace, the deep sea, and the polar regions. This is because these areas are experiencing increased human activity, evolving physical environments, and some, such as the Arctic, are not governed by region-specific and binding international treaties and as such might be seen by some as domains requiring adjustments in their governance arrangements.

The Chinese government recognises the Arctic region as a global space where extra-regional states have certain rights and privileges under international law. It has also indicated that it wants to contribute constructively to the development of rules and regulations for the Arctic while also upholding the existing regional governance system based on the UN Charter and UNCLOS.[3] However, some within the Chinese Party-state system acknowledge inefficiencies in the current Arctic governance system and see room for improvement. This chapter explores the Chinese official and academic discourses regarding Arctic governance, examining its challenges and the normative solutions Chinese analysts propose to address them. It is divided into three sections. The first section provides an overview of the current Arctic governance system, highlighting the role of Arctic states, the Arctic Council, and the numerous challenges faced by the region. The second section outlines the official PRC position on regional governance. This is followed by an analysis of Chinese academic discourses on Arctic governance, focusing on identified inefficiencies, challenges, shortcomings, and solutions proposed by academics and analysts. The third section explores the PRC's bilateral relations with Arctic

DOI: 10.4324/9781003295112-5

states as the extent of China's participation in regional affairs is influenced by how these states perceive Chinese activities in the Arctic.

Arctic Regional Governance

The current Arctic governance system emerged as a result of significant changes that occurred in the region in the 1990s, namely, the disintegration of the Soviet empire, the end of the Cold War, and the subsequent demilitarisation and de-escalation in the region.[4] While some international observers compare the Arctic to Antarctica and note the absence of a single regional treaty or international organisation with the mandate to impose and implement legally binding rules and regulations, Young points out that this does not imply that the region is devoid of governance.[5] In fact, there is a number of international regimes, treaties, and agreements, such as the United Nations Convention on the Law of the Sea (UNCLOS), the Svalbard Treaty, the International Convention for the Prevention of Pollution from Ships (MARPOL), and the International Maritime Organization's Polar Code, to which the Arctic is a subject to and which states that want to operate in the Arctic must take into consideration. Consequently, Arctic governance has evolved into what analysts describe as "a mosaic of issue-specific arrangements"[6] or a collection of formal and informal provisions that span across various areas and operate at different levels (international, regional, and national) creating a form of multilevel governance.[7]

A key feature of this governance system is the prominent role played by the eight Arctic governments – those that have territories within the Arctic – Russia, Finland, Sweden, Norway, Denmark, Iceland, Canada, and the US. Currently, their interests are paramount in addressing governance needs in the region.[8] Due to their geographical location, these states perceive themselves as stewards of the Arctic.[9] This perception was echoed by former US State Secretary Mike Pompeo in 2019 when he declared that there were only Arctic and non-Arctic states in the region, rendering any further distinctions irrelevant.[10] While occasional differences may arise among the Arctic states regarding control and authority in the region, they generally maintain a united front regarding their primacy in addressing issues of Arctic governance.[11]

To discuss regional issues, these states convene within the Arctic Council (AC). The AC was formally established in 1996 through the Ottawa Declaration, which marked the culmination of a process initiated in the late 1980s when Mikhail Gorbachev, then General Secretary of the Communist Party of the Soviet Union, envisioned the Arctic as a zone of peace. Soon afterward, Finnish representatives brought together officials from circumpolar Arctic governments and launched the Arctic Environmental Protection Strategy (AEPS) in 1991. Under the umbrella of the AEPS, representatives from Arctic states began regular meetings, established various working groups, and invited Indigenous groups and non-Arctic states to participate, leading to the formation of the AC as a new forum.[12]

The AC consists of the eight Arctic states as its members. Additionally, an essential part of the AC is the inclusion of the so-called Permanent Participants (PP), which are six organisations representing Arctic Indigenous Peoples such as the Saami Council.[13] The Council conducts its work through six working groups, several task forces, and one expert group, covering various fields ranging from climate change to emergency responses.[14] Decision-making authority resides solely with the Arctic states, and the PP must be fully consulted before any decisions are made. Although not an international organisation, the AC has a permanent secretariat in Tromsø, Norway, established in 2011 and operational since 2013. Each Arctic state assumes the chairmanship of the AC for a two-year period and appoints a Senior Arctic Official to represent its interests within the Council. Additionally, every two years, the AC holds a high-level ministerial meeting attended by foreign ministers and secretaries of state of the Arctic states.[15]

In addition to member states and the PP, the AC involves a wide range of observers. As of 2023, there were a total of 38 observers, including non-Arctic states (both European and Asian), intergovernmental and interparliamentary organisations (such as the World Meteorological Organisation and the Standing Committee of the Parliamentarians of the Arctic Region), as well as various non-governmental organisations (e.g., the Association of World Reindeer Herders and the University of the Arctic). The primary role of observers is to observe the work of the Council. They do not possess voting rights, as decision-making at all levels of the Council is exclusively reserved for member states. However, they can contribute to the work of the AC through participation in the various working groups and task forces, and they may propose projects through a member state or a PP, as long as the funding they provide does not exceed that of a member state. Observers are also allowed to submit written statements at the high-level ministerial meetings of the AC.[16]

Nevertheless, the AC does have its limitations. As mentioned previously, it is a regional forum and not an international organisation with a legal body. Consequently, it lacks its own programming budget and relies on contributions from member states and other sources for funding. Furthermore, the AC does not have the authority to enforce its guidelines and recommendations. In recent years, three legally binding agreements have been negotiated among the Arctic states under the auspices of the AC (on search and rescue, oil pollution, and scientific cooperation). However, the AC served solely as a negotiation platform, and the enforcement of these agreements rests in the hands of the Arctic states. It is also worth noting that the AC's thematic focus primarily revolves around environmental protection and sustainable development, explicitly excluding issues pertaining to military security.[17]

Finally, regional governance in the Arctic was based on the assumption that the region was shielded from broader geopolitical events,[18] making it a place of supposed functional cooperation, low tensions, and peaceful co-existence.[19] Furthermore, Heininen, Exner-Pirot, and Plouffe argue that several other characteristics of regional governance make the Arctic appear unique. They highlight the importance placed on consultations, engagement, and consensus-building in governance deliberations, the involvement of Indigenous groups and local governments, as well as the significant role of epistemic communities in policy development.[20] These factors contributed to the perception, since the end of the

Cold War, that the Arctic was an exceptional space distinct from other regions, where cooperation and an emphasis on environmental protection and sustainable development prevailed.[21]

Yet ongoing environmental and political changes in the region challenge the notion of Arctic exceptionalism and the broad contours of regional governance, as outlined previously.[22] The most notable are the Arctic environmental changes, in particular the ones related to rising temperatures in the region. These changes have far-reaching implications, including the disappearance of Arctic sea ice, the melting of permafrost, increased wildfires, and potential alterations to established weather patterns.[23] Beyond environmental concerns, the changing physical environment of the Arctic also opens up opportunities for greater access to natural resources such as hydrocarbon and rare-earth elements.[24] These developments, however, extend far beyond the Arctic region, with Arctic environmental changes linked to the increased occurrence of extreme weather events in the mid-latitudes of the Northern Hemisphere. Additionally, the region's economic potential has attracted the interest of extra-regional governments and enterprises from across Europe and Asia. All of these factors indicate that Arctic issues and opportunities are merging into a global environmental and economic agenda,[25] challenging the dominant position of the Arctic states, and making it increasingly difficult to define who qualifies as an Arctic stakeholder.[26] Concurrently, there is a growing number of multilateral venues, such as the Arctic Frontiers in Norway, that extra-regional states and actors can utilise to exchange views and ideas about regional developments, including military and security issues, thus potentially supplementing the role of the AC in regional governance.[27]

However, the re-militarisation of the Arctic region and the spillover effects from the intensification of global great power competition pose the most significant challenges to the regional governance system. Since 2007, when Russia resumed the flights of its long-range strategic bombers over the Arctic, Moscow has been constructing and reopening new military bases in its Arctic zone, raising concerns among other Arctic governments.[28] The annexation of Crimea in 2014 further exacerbated tensions between Russia and the West. While cooperation within the AC remained relatively unaffected, the first Ukraine crisis led to an increased distrust of Russia in the West, the implementation of economic sanctions against the Russian energy sector, and the breakdown of regional military exchanges, including the cancellation of the annual Arctic Chiefs of Defence meetings.[29] As a consequence, the US began to pay more strategic attention to the European Arctic, upgrading its military facilities in Greenland, resuming operations on its former base in Iceland, and engaging in more military exercises with its European allies over the Arctic Circle.[30] This has led some observers to conclude that the region exhibits features of an Arctic security dilemma.[31] However, regional cooperative engagements suffered a complete breakdown after Russia's invasion of Ukraine in early 2022. All exchanges between Russia and the other seven Arctic states were halted, and the work of the AC was suspended.[32] Since then, there has been no significant progress in Russia-West Arctic relations. Although Norway successfully assumed the AC Chairmanship from Russia in May 2023, several Arctic states, including the US, have indicated that near-term cooperation with Russia in the Arctic appears unlikely under current conditions.[33]

The PRC and Arctic Governance: Official and Academic Discourses

The PRC's official position regarding Arctic governance, as outlined in its Arctic White Paper, is that it "upholds the current Arctic governance system."[34] The country's participation in Arctic governance primarily takes place through global and regional level treaties, mechanisms, and organisations. For example, China is a signatory of the Svalbard Treaty and UNCLOS, which it regards as the primary instruments legitimising its participation in regional affairs and governance.[35] China is also an observer to the AC. It became an ad hoc observer in 2007 (after submitting its official application in December 2006) and a full observer in 2013. According to Lanteigne, China gradually recognised that as the Arctic gained prominence in the international arena due to environmental changes, the Council could provide valuable "club goods" (that is, goods retained by members of a particular regime that are denied to outsiders), such as information gathering about the region and the opportunity to be at the table with regional decision-makers.[36] Presently, according to documents submitted to the AC, China recognises the AC as "the main intergovernmental forum on issues regarding the environment and sustainable development of the Arctic" and values its positive contributions to Arctic affairs.[37] The PRC has also appointed a Special Representative for Arctic Affairs from the MFA to represent China during political gatherings of the AC, including high-level ministerial meetings. In addition to these meetings, China also sends researchers and experts from various parts of the Party-state system, including the Ministry of Transport, the China Meteorological Administration, the CAS, the National Marine Environmental Monitoring Centre, and Tongji University to participate in the AC's working groups.[38]

Regarding the PRC's engagement with Arctic Indigenous Peoples, these seem to be limited compared to other Asian observers of the AC, such as Singapore, which actively engages with these peoples through various programs such as study visits.[39] There were previous reports of Chinese officials, particularly those assigned to diplomatic missions in the Nordic states, attempting to reach out to the Saami people and the Indigenous Peoples Secretariat in 2011 to seek their perspectives on Arctic affairs.[40] More recently, in 2016, China contributed to the Indigenous Peoples Secretariat's project – A story map of Indigenous Peoples and the AC – to commemorate the 20th anniversary of the AC.[41] However, there do not seem to be any other significant developments. This lack of interaction could be attributed to the PRC's preferences for dealing with actors that represent the entire state rather than non-governmental entities. Alternatively, Bennett notes that China, as well as Japan and India, may be hesitant to support greater engagement and Indigenous authority in the Arctic to avoid encouraging similar movements for support and legal recognition from Indigenous Peoples within their own borders (the Chinese Party-state officially recognises 55 ethnic minorities in addition to the dominant Han).[42] However, some Chinese researchers have suggested that the PRC could reconsider this stance. For example, Peng and Lu propose that establishing good relations with indigenous populations could be beneficial for China's future investment projects in the Arctic, as potentially profitable mines are located on lands traditionally owned by native communities.[43]

While publicly declaring that the PRC upholds the current governance system in the Arctic, there are indications and subtle hints from Chinese officials and documents that suggest the possibility of enhancing Arctic governance, including regional economic development. For example, the PRC's Arctic White Paper notes that China "is committed to improving and complementing the Arctic governance regime" and building "a just, reasonable, and well-organized Arctic governance system."[44] Similarly, its Special Representative for Arctic Affairs argued that "through diversified cooperation, a better institutional system could be put in place for the sustainable development of the Arctic."[45] Moreover, since designating itself as an important Arctic stakeholder, the PRC now sees itself as an Arctic affairs active participant, builder, and contributor,[46] with the capability to provide Chinese wisdom and Chinese strength (中国智慧和中国力量) for the improvement of Arctic sustainable development, according to Chinese analysts.[47]

However, PRC officials have also indicated that China will "not overstep" (不越位) its role as a non-Arctic state, while at the same time making it abundantly clear that it will "not be absent" (不缺位) from regional affairs.[48] Kong Xuanyou from the Chinese MFA explained that the PRC will not intervene in Arctic affairs related to matters between Arctic states, that it will participate in regional affairs in accordance with international laws and instruct Chinese entities and individuals to conduct Arctic-related activities in compliance with these laws and the domestic laws of Arctic states. Nevertheless, Kong emphasised that as an important stakeholder in Arctic affairs, the PRC strives to play an active and constructive role in Arctic cross-regional and global issues.[49] The second point is particularly important from the PRC's perspective, as China does not want to be excluded from regional affairs. The PRC maintains that regional Arctic governance "requires the participation and contribution of all stakeholders,"[50] and it advocates for the respect and preservation of the rights of non-Arctic states, including scientific research, navigation, fishing, and resource exploration.[51] This is because the PRC aims to avoid a scenario, described by Lanteigne as "the blueberry-pie scenario," in which the region is divided among the eight Arctic states, leaving extra-regional actors, such as the PRC, with limited access to the region.[52]

Similarly, many individuals within the Chinese Party-state system perceive regional Arctic affairs, particularly the AC, as restrictive in determining who can participate and have a voice in regional affairs. Xu Hong from the Chinese MFA clearly articulated this position, noting that observer states in the AC are subject to numerous restrictions in their actual participation in the Council's work. While conceding that there have been some changes in observer engagements within the Council in recent years, including arrangements such as special observer meetings, Xu nevertheless concluded that the level of participation has not improved significantly.[53] Consequently, Xu notes that extra-regional states have developed "centrifugal tendencies" (离心倾向), that is, these states started seeking alternative venues and platforms to share their ideas, opinions, and concerns regarding Arctic affairs.[54] The Arctic Circle Assembly in Iceland has emerged as such a venue for the PRC. Since its inception in 2013, China has regularly attended this conference, sending large delegations consisting of government representatives, SOEs, scientists, and academics. During the 2022 session, the PRC's Special Representative for Arctic Affairs affirmed China's long-standing affiliation with the Arctic Circle Assembly, calling it "an important and unique platform for the discussion of Arctic affairs."[55] If the AC remains in a state of uncertainty, these alternative platforms, with robust participation from various regional and extra-regional stakeholders, will gain even greater importance in the Arctic governance system, providing the PRC with increased opportunities to express its views on regional affairs.

However, beyond the official government discourse on regional affairs, Chinese analysts and academics are more vocal in their criticisms of the current state of Arctic governance. Many Chinese Arctic experts view the existing Arctic governance system as inadequate in addressing the region's most pressing needs, such as environmental changes and economic development. They consider it exclusive, with limited input from extra-regional states, and identify several contradictions that hinder its effectiveness, suggesting the need for new and innovative governance approaches. Zhang Yao, the Director of the Centre for Maritime and Polar Studies at the SIIS, highlights two bottlenecks in Arctic governance and cooperation. He argues that the Arctic "lacks an effective, authoritative, and representative international governance mechanism," while Arctic states maintain a conservative and inward-looking approach to governance, making it difficult for extra-regional states to express their concerns.[56] Sun and Wu from the Ocean University of China identify several contradictions in their analysis of Arctic governance, including the global nature of issues related to the Arctic and the exclusionary tendencies of Arctic states toward non-Arctic states, the balance between the need for Arctic economic development and environmental protection, the lack of compliance mechanisms in regional affairs and the urgency for regional governance, and finally, the need for regional cooperation and the influence of global geopolitics on Arctic governance.[57] Luo and Zha from Central South University in Changsha go further, suggesting that Arctic governance has undergone an "involution" (内卷化), characterised by a lack of openness, institutional defects, and a weak response capacity within regional governance mechanisms in the Arctic.[58]

A particularly prominent criticism within Chinese analytical and academic circles regarding Arctic governance and regional affairs is the dominant role of the eight Arctic states in the region and what Chinese observers perceive as exclusionary and restrictive behaviour toward non-Arctic actors in Arctic development and governance mechanisms. Deng, for example, argues that areas in the Arctic beyond the national jurisdiction of states (referred to as the Arctic commons) face the challenge of "privatisation" (私域化) by Arctic states, which has implications for non-Arctic states and their potential interests in the region. Deng identifies several manifestations of this process. For instance, he notes that Russia and Canada are asserting varying degrees of sovereignty over Arctic shipping lanes, imposing mandatory application and reporting requirements for foreign commercial vessels.

He also contends that Arctic coastal states are continuously expanding their continental shelf claims and as such reducing the size of the international seabed. Moreover, he suggests that some Arctic states seek to regulate activities that are permitted under international law, such as scientific research and ocean surveys, through the expansion of their continental shelf. Deng further argues that Arctic states leverage their geographical advantage in the region to establish dominant institutional arrangements. As an example, he points to the recently concluded Central Arctic Ocean Fisheries Agreement, which involved negotiations between five Arctic coastal states (the US, Canada, Russia, Denmark, and Norway) and five other stakeholders (China, Iceland, Japan, Korea, and the European Union). The agreement stipulates a prohibition on commercial fishing activities in the Central Arctic Ocean for at least 16 years.[59] However, Deng sees this as another indication of the commanding position of Arctic states in the region. He notes that Arctic coastal states, through multiple rounds of negotiations since 2007, have exploited their geographical advantage to shape rules while disregarding the interests of the international community in the high seas of the Arctic Ocean. Only later, in order to address the legitimacy deficit at the operational level, did the Arctic states invite other stakeholders to participate in the negotiations. Deng contends that this selective inclusion of other Arctic Ocean fisheries stakeholders merely underscores the dominant position of the Arctic coastal states in regional affairs and their desire to exert control over regional developments. Lastly, Deng discusses the Svalbard Treaty, noting that the signatories of the Treaty are concerned that Norway, through its own interpretation, may extend its sovereignty claims beyond fisheries into areas such as hydrocarbon development, scientific research, and seabed exploration.[60]

In a similar fashion, many Chinese analysts highlight what they perceive as the exclusiveness or exclusivity (排他性) of the Arctic governance structure. This is particularly relevant for the AC and the dominant position held by the eight Arctic governments within it. According to Chinese academic writings, these governments aim to exclude non-Arctic states from participating in regional affairs through bilateral and multilateral arrangements, a practice that can be translated as "internal cooperation, external exclusion" (内部合作，外部排他).[61] Chinese analysts specifically criticise the criteria for admitting new observers to the AC, which were first issued in 2011 at the ministerial meeting in Nuuk, Greenland, and later adopted in the 2013 Arctic Council Observer Manual. These criteria include, among others, the recognition of Arctic states' sovereignty, sovereign rights, and jurisdiction in the Arctic; the acknowledgement that an extensive legal framework, notably the Law of the Sea, applies to the Arctic Ocean, and provides a foundation for responsible management, and the respect for the values, interests, culture, and traditions of Arctic Indigenous Peoples and other Arctic inhabitants.[62] The "three recognitions" – sovereignty, sovereign rights, and jurisdiction – seem to generate the most controversy. For example, Guo Peiqing from the Ocean University of China finds these criteria illogical and problematic. He questions how non-Arctic states can be required to recognise Arctic states' sovereign rights in the region when Arctic states themselves have not resolved these issues. He points to existing maritime delimitation disputes (such as those between the US and Canada in the

Beaufort Sea), the delimitation of the outer continental shelf, the status of waters around the Svalbard archipelago, and the applicability of the Svalbard Treaty. Guo concludes that with the adoption of these criteria, non-Arctic states were compelled "to surrender important user-state entitlements to the Arctic Ocean" and bringing "non-Arctic states more obligations than rights and benefits" while making the observer status in the AC "a suboptimal choice for non-Arctic states."[63] This has led some within Chinese academia to conclude that a version of the Monroe Doctrine (originally adopted by the US in the 19th century to keep European powers out of the Western Hemisphere) was at work in the Arctic region. According to Xiao Yang, the AC observer criteria resemble the Monroe Declaration because they aim to exclude non-Arctic states from Arctic governance, ensuring that observer states cannot become full members of the AC, preserving the dominant position of Arctic states, and in general bringing more obligations than rights to non-Arctic observer states.[64] Likewise, Pan Min argues that the "three recognitions" represent an attempt by Arctic states to restrain non-Arctic states and effectively declare to the world that "only Arctic states have the right to the Arctic."[65]

Another development within Arctic affairs that concerns Chinese Arctic specialists, analysts, and observers is what they perceive as the emerging remilitarisation of the region and the nascent great power competition (or games 大国博弈) between regional powers, particularly the US and Russia.[66] Given the recent revival of Russian Arctic military power, the strategic return of the US to the region, increases in regional military exercises, air patrols, and reconnaissance missions, and growing interest from countries like India and South Korea, some Chinese analysts view the Arctic as becoming a "new battlefield" in the strategic competition between Arctic states, near-Arctic states, and important Arctic stakeholders.[67] From the Chinese perspective, the US is seen as the main catalyst for these developments. With its strategic return to the Arctic and the publication of its new Arctic strategy, which focuses, among other things, on issues of regional security, Chinese analysts believe that the US is not only seeking to expand its power and presence in the region but also to maintain its leading position in the Arctic, potentially working with its regional allies to obstruct the participation of "certain" countries in Arctic affairs.[68] This is then contributing to increased regional tensions reminiscent of the Cold War mentality of bloc confrontations.[69] Furthermore, some Chinese academics argue that the current governance system is ill-equipped to address hard security issues. Jiang Yin'an from the China Institute of International Studies, a think tank under the Chinese MFA, notes deficiencies in security mechanisms in the Arctic, leading to a lack of effective channels for regional military communication, exchange, and cooperation.[70] The deteriorating security environment in the Arctic has negative implications for China's regional engagement, according to Chinese specialists. Zuo Pengfei, for example, argues that the intensification of regional military competition, particularly the American redeployment of military forces, significantly reduces the PRC's strategic space in the Arctic and undermines its legitimate security interests.[71] A more restrictive Arctic security environment will also subject the activities of Chinese commercial actors, many of which have close ties to the Party-state, to increased scrutiny. As a result, the availability of investment projects for these actors, especially in parts of the Arctic controlled by the US and its allies, could decrease.

Chinese Solutions

To address the perceived inefficiencies of the Arctic governance system and accommodate extra-regional interests in the Arctic, Chinese academics and analysts propose various adjustments, reforms, and innovations to regional Arctic governance.[72] While there is a wide spectrum of ideas on what to improve, they all share the common goal of establishing a global, more inclusive governance system that would take into consideration the interests of stakeholders beyond the eight Arctic governments. For example, Yang Jian from the SIIS, cautiously suggests that "there is a need to encourage constructive engagement on the part of many actors, without distorting the architecture of the Arctic Council or undermining its unique features."[73] Pan Min and Xu Lining argue that international forums such as the Arctic Science Ministerial are better positioned to address regional challenges than the AC. This is because the Arctic Science Ministerial represents a global, equal, and inclusive vision of Arctic governance that includes Arctic and non-Arctic states, indigenous representatives, and international organisations. The science ministerial further highlights the global impact of Arctic changes and stresses the need for global action to combat them. More importantly, it underlines the scientific contributions of non-Arctic states and links decision-makers directly with the science community. This broad and global inclusion of non-Arctic states can help overcome the perceived Arctic governance Monroe Doctrine.[74] Additionally, Li and Liu from Dalian Maritime University propose the establishment of a Greater Arctic Mechanism (大北极机制) that goes beyond the eight Arctic governments and includes the interests of states from across the Northern Hemisphere to maintain security and stability in the Arctic region, particularly in relation to the opening of Arctic shipping routes.[75]

Reflecting on the diplomatic proposals advanced by General Secretary Xi, some Chinese analysts believe that the application of the concept of the Community of Shared Future for Mankind could be a plausible way forward to improve Arctic governance and address regional issues.[76] Xi's statement in 2017 at the UN Office in Geneva emphasised that the strategic new frontiers, including the polar regions, should be areas of cooperation "rather than a wrestling ground for competition," based on principles of peace, sovereignty, inclusiveness, and shared governance.[77] Bai Jiayu from the Ocean University of China sees the Community of Shared Future for Mankind as a new approach to international relations and the governance of the Arctic Commons. She argues that the Community, rooted in traditional Chinese culture and Marxist ideas of a social community, surpasses Western conceptions and theories of international relations, opposes hegemony, advocates the establishment of a new type of major power relations, and emphasises cooperation and tolerance between cultures. By promoting win-win cooperation, inclusiveness, mutual understanding, and a global perspective on the governance of Arctic commons as the necessary preconditions, the Community can help address regional dilemmas and governance deficiencies, such as limited national interests, the impacts of great power competition, and the fragmented nature of regional governance mechanisms.[78]

In a similar vein, Li et al. observe that the establishment of an Arctic Community of Shared Future for Mankind represents the most effective approach to addressing Arctic issues from the perspective of sustainable development. This is because, as they argue, the notion of a Community of Shared Future for Mankind transcends the narrow pursuit of national interests by considering the interests of the broader international community and prioritising cooperation over competition.[79] Taking it a step further, Xu Guangmiao from Wuhan University suggests that an Arctic Community of Shared Future for Mankind is not only an integral part of the broader Community of Shared Future for Mankind but also represents a grand vision for the evolution of the Arctic order.[80] The idea of an inclusive Arctic based on common benefits and shared interests, as espoused in the Community of Shared Future for Mankind approach to regional governance, exemplifies Chinese-specific solutions and wisdom, which the CCP seems to be so keen on providing to a world grappling with various governance deficits. Chinese experts assert that the Polar Silk Road could serve as the platform or vehicle through which China promotes the building of an Arctic Community of Shared Future for Mankind.[81]

The concept of the PSR emerged from Sino-Russian considerations on developing the shipping potential of the NSR in the mid-2010s.[82] According to the Chinese Arctic White Paper, the PRC government encourages Chinese companies to conduct commercial trial voyages using Arctic shipping routes and participate in infrastructure development along these routes to support the PSR's development.[83] Sun highlights that this could enhance economic connectivity in the region and further integrate the Arctic with the international community.[84] However, as Lanteigne points out, the PSR's original scope has now expanded beyond Arctic shipping to encompass bilateral and multilateral initiatives covering energy, infrastructure and science cooperation, digital connectivity, and tourism.[85] Yang articulates the vision of the PSR as a vehicle for Arctic multilateralism that could be "a new platform for policy coordination and science, industrial, social collaboration among various countries," including Arctic and non-Arctic states.[86] From the Chinese perspective, prioritising Arctic multilateralism and international cooperation through initiatives such as the PSR is crucial. This approach allows China to promote the idea of an inclusive Arctic that considers the interests of all stakeholders, both within and outside the region, thus preventing the privatisation of Arctic commons (as described previously).[87]

Chinese analysts also recognise that in order for China to effectively promote its ideas and visions for Arctic governance and regional sustainable development, it needs to enhance its Arctic discourse power.[88] The development of Arctic shipping routes and Chinese investments in resource extraction projects, as part of the PSR construction, provide China with a tangible material foundation to bolster its discourse power. Additionally, conducting significant science and high-impact research projects in the Arctic is seen as another avenue for China to increase its discourse power. Yang notes that the scientific and technological capabilities of established polar states have played a crucial role in enhancing their discourse power in polar affairs. Therefore, he urges the Chinese government to focus on further enhancing China's polar research and technological capabilities.[89] An equally important factor that could enhance Chinese Arctic discourse power is to nurture competent human capital, particularly highly trained and skilled polar scientists. These experts would not only provide relevant information to the Chinese leadership on regional issues and developments but, for example, through their placement in polar-focused international organisations, also be in a position to influence regional policy development through agenda-setting processes.[90]

The focus on Chinese Arctic scientists, research projects, and international scientific cooperation in the context of increasing Chinese influence in regional affairs and governance is relevant when considering science and scientific cooperation as an instrument that a state can utilise in its foreign diplomatic strategies, including science diplomacy, to achieve favourable international outcomes. One aspect often discussed in science diplomacy is the potential of international scientific cooperation to bridge gaps and enhance communication between states, encapsulated in the term "science for diplomacy."[91] While most scientific engagements within the realm of science for diplomacy occur between epistemic communities, it is crucial to recognise the significant role that governments and decision-makers play in science diplomacy, linking it to state policy.[92] The nature of the PRC as a Party-state in which the governing organs of the CCP are spread throughout the government-funded institutions only accentuates the notion of science diplomacy as part of state policy.

In recent years, Chinese scientists, research organisations, think tanks, and universities have engaged in what can be termed "track-two diplomacy" – an inclusive, multilateral, and often informal approach to diplomacy conducted by various actors, such as academics, as opposed to "track-one diplomacy" conducted by official state representatives.[93] For example, Chinese Arctic specialists regularly engage their Russian counterparts through the China-Russia Arctic Forum established in 2012. It is through such track-two diplomacy platforms that Chinese scientists and researchers interact with their foreign counterparts and communicate Chinese perspectives on regional affairs. According to Su and Wu, Chinese institutional participation in track-two diplomacy strengthens exchanges between China and Arctic countries, dispels foreign scholars' misunderstandings about China's involvement in Arctic affairs, enhances China's discourse power, and expands its interests in the Arctic region.[94] Cheng Xiao adds that Chinese research institutes and universities can play an important role in China's Arctic strategy due to their "unofficial" character, which makes it easier for them to engage in international exchanges and helps counter negative perceptions of China in the Arctic.[95]

Finally, from the perspective of increasing Chinese influence and position in Arctic governance, it is important to note that institutional and physical presence in the region through research stations and other scientific infrastructure can be perceived as geopolitically motivated. For example, Pedersen points to the use of national and cultural symbols at research facilities in Svalbard leased by foreign research institutions. He maintains that such indications of national presence, or the strategic presence of a state actor rather than individual nationals such as scientists (PRIC installed two stone "guardian lions" at the entrance to the Chinese Arctic Yellow River Station as symbols of Chineseness), signify a form of national posturing by foreign governments to establish a foothold in the Arctic. Drawing on naval strategy terminology, Pedersen likens this to "showing the flag," which is an element of strategic presence – a tool of statecraft and instrument of influence.[96]

Considering the broader narratives and criticism of Arctic governance by the Chinese government and academia, including the perceived exclusionary structures of the AC, as well as Chinese proposals to establish a Community of Shared Future for Mankind-like regional governance mechanism, it becomes clear that the PRC desires a more inclusive regional order that involves extra-regional actors and accommodates the interests of the broader international community. This is particularly evident in the PRC's framing of Arctic environmental issues as a common challenge for the international community that can only be addressed with the participation of all states. In essence, the PRC aims to internationalise the Arctic while simultaneously positioning itself to contribute to addressing regional issues that the current governance regime cannot effectively handle due to its inefficiencies and biases.[97] According to Lanteigne, the PRC has effectively become a norm entrepreneur – a state "which invites other actors to participate in a re-evaluation of a particular concept or policy . . . through dialogue in hope of creating a different norm, one more acceptable to the entrepreneur."[98] In the Arctic context, "China is selling the idea of an Arctic developed and structured via partnerships and regimes between Arctic and non-Arctic actors for mutual benefit, while ensuring that China's own interests in the region can be enhanced."[99] In this regard, high-profile international conferences, such as the Arctic Circle Assembly in Iceland or the Arctic Frontiers in Norway, can serve as convenient platforms for the PRC to promote its ideas and visions of regional governance and sustainable development. In 2016, China's Special Representative for Arctic Affairs declared that "the Arctic Circle has become the most dynamic and viable multilateral platform for comprehensive discussions on Arctic issues."[100] Zhang Jiajia from Wuhan University suggests that the PRC must leverage these venues and their role in Arctic governance to challenge the discourse hegemony of Arctic states.[101] With China's more assertive stance in international affairs and its willingness to assume the role of a norm-maker in global governance under Xi, China will likely seek a greater role in Arctic affairs in the near future. In this regard, Hong Nong, the Executive Director of the Institute of China-America Studies, confidently concludes that recent Chinese Arctic activities and the publication of China's Arctic White Paper provide "evidence of China's willingness and ability to be more than a mere observer to the Arctic Council."[102]

PRC's Bilateral Relations With Arctic States

China's scientific and technical capabilities and political views on the Arctic are key factors in understanding its engagement in regional governance. However, an equally important factor that will influence the form and degree of its participation in Arctic governance is the evolving relationship between China and the Arctic states. These states, by virtue of their geography and membership status at the AC, and to the discomfort of the PRC, can either facilitate or hinder Chinese presence in the region at both the national and subnational levels. For example, China's Arctic research, to a certain degree, depends on the willingness of Arctic states to accommodate Chinese scientific infrastructure on their sovereign territory. As a result, China must carefully consider its bilateral relations with regional actors, especially as the Arctic security environment becomes more restrictive. This changing Arctic reality and emerging threat perceptions influence how China is perceived in various Arctic state capitals. The following section examines Sino-Russian relations, Sino-American relations, and Sino-Nordic relations, presenting a spectrum of possibilities for China, ranging from cooperation to potential confrontation.

The Arctic in Sino-Russian Relations

The Russian Arctic zone encompasses approximately half of the Arctic coast and holds immense resource wealth, including natural gas, oil, precious metals, gems, and rare-earth ore deposits.[103] It is also home to a significant portion of the Russian Navy and its strategic forces. Furthermore, the Russian leadership has high aspirations for the development and utilisation of the NSR as an economically viable shipping route. Russian Arctic policy, outlined in the 2020 document – *Foundations of the Russian Federation State Policy in the Arctic for the Period up to 2035*, emphasises the need to ensure sovereignty and territorial integrity, preserve the region as a territory of peace, stability, and mutually beneficial partnership, improve the well-being of the Arctic population, develop the region's resource base and the NSR as a competitive national transportation passage in the global market, protect the environment, and preserve the traditional way of life of Arctic Indigenous Peoples.[104] Furthermore, amid deteriorating relations with the West and the resultant departure of Western companies from the Russian market, President Putin has expressed openness to involving extra-regional states and associations in Arctic cooperation.[105] This presents an opportunity for Chinese actors to expand their involvement in the Russian Arctic.

Since the mid-2010s, the Arctic has been emerging as an area of pragmatic engagement and cooperation between China and Russia. This occurs amid the seemingly ever expanding Sino-Russian political, economic, as well as military partnership.[106] General Secretary Xi Jinping has held over 40 meetings with President Vladimir Putin, and the two countries maintain regular government-to-government meetings (the year 2022 witnessed their 27th iteration).[107] They also provide diplomatic support to each other on issues deemed crucial to their national interests, including on matters related to Taiwan or the war in Ukraine.[108] For example, a joint statement issued after Xi visited Moscow in 2023 notes that Russia "recognizes Taiwan as an inalienable part of China's territory," thus supporting the PRC's claims over the island.[109] In terms of economic relations, bilateral trade (in energy, raw materials, agricultural products, and more) is steadily growing, with a potential to exceed a record-breaking USD 200 billion.[110] While both countries clarify that their military ties do not constitute a security alliance,[111] their armed forces conduct joint military exercises, including army and navy drills and strategic bomber patrols.[112] This cooperation also extends to more sensitive areas. For example, in 2019, President Putin announced that Russia was assisting the PRC in developing a missile early warning system which, according to Chinese analysts, will significantly enhance the PRC's anti-missile and sky-monitoring capabilities.[113]

At the political level, the Arctic's relative importance in bilateral Sino-Russian relations is evidenced by its inclusion in various government meetings and agreements. The Chinese MFA and its Russian counterpart regularly exchange views on Arctic affairs, including the work of the AC, through the Sino-Russian Arctic Affairs dialogue.[114] The 2019 *Sino-Russian Joint Statement on Developing Comprehensive Strategic Partnership of Coordination in the New Era* also highlights Arctic engagement as a key area of cooperation. It stipulates that both countries will promote cooperation in the areas of Arctic resource and infrastructure development, the utilisation of Arctic shipping routes, tourism, scientific research, and environmental protection.[115] Additionally, high-level meetings between General Secretary Xi and President Putin have reportedly addressed issues related to Arctic energy and transportation.[116]

As outlined in Chapter 4, the PRC's commercial involvement in the Arctic can contribute to its economic and technological security, particularly in energy supply, shipping, engineering, and shipbuilding. However, with Russia's willingness to include more extra-regional actors in the development of the Russian Arctic, new opportunities are also emerging. Chinese enterprises could participate in prospecting for and extracting natural resources in the Russian Arctic zone, drawing on their existing expertise. For instance, CNOOC's subsidiary, China Oilfield Services Limited (COSL), has previously conducted seismic mapping in the Barents Sea[117] and drilling operations in the Kara Sea in Russian Arctic waters, resulting in the discovery of significant natural gas deposits.[118] Similarly, in 2023, Russian Titanium Resource, a company specialising in titanium and quartz raw materials, announced a partnership with the China Communications Construction Company (CCCC), a major SOE and contractor for Chinese BRI projects, to develop titanium ore deposits in the Russian Arctic's Komi Republic.[119] Another potential avenue for cooperation involves satellite data and remote sensing. Rosatom, the main operator of Russia's fleet of nuclear icebreakers, recently expressed interest in partnering with a Chinese entity to obtain space-based data for improved prediction of sea-ice conditions and navigation on the NSR.[120]

Chinese analysts also believe that a massive economic potential lies in the development of infrastructure, such as ports, railways, and highways, along the NSR.[121] The Chongyang Institute for Financial Studies at Renmin University, an influential think tank advising various Party-state institutions, including the Chinese MFA and the CCP's International Department,[122] published a report in 2018 titled *Going to Europe, head North: Research on the establishment of joint Sino-Russia Polar Silk Road pivot ports*. This report evaluated several ports along the NSR and identified Murmansk, Arkhangelsk, Sabetta, Tiksi, and Uelen as potential pivot ports (支点港口) for the PSR. The selection was based on factors such as geographic location, development potential, natural conditions, demographic factors, and the basis for Sino-Russian cooperation. The report argues that these pivot ports, envisioned as a cargo distribution, ship supply, and maintenance network, will allow China and Russia to fully unlock the economic potential of the PSR.[123] In this regard, Chinese enterprises have previously expressed interest in port projects along the NSR, particularly in Murmansk and Arkhangelsk.[124] Furthermore, there have been reports of Chinese investors, such as the Poly Group, considering funding the development of the Belkomur railway – a 1,000 km infrastructure project connecting western Siberia to the port of Arkhangelsk.[125] Likewise, the aforementioned titanium ore extraction project involving CCCC also includes plans for the Indiga deep-water port and the construction of a railway linking the port to the regional mining cluster in the Komi Republic, enabling the transportation of cargo to the NSR.[126] Additionally, CCCC was awarded a contract to construct an LNG transhipment terminal in Kamchatka, in the Russian Far East, that will handle LNG shipments from the Arctic to Asia.[127]

In addition to economic opportunities, the vast Russian Arctic zone and the considerable Russian Arctic scientific expertise offer further prospects for research-oriented elements of the Chinese Party-state to enhance their understanding of the region. Over the past decade, Chinese and Russian scientists and experts have embarked on several joint Arctic scientific expeditions, established collaborative research centres, and engaged in bilateral academic exchanges. For instance, in 2016, 11 Chinese researchers from the First, Second, and Third Institutes of Oceanography, the PRIC, Ocean University of China, and Xiamen University joined their Russian counterparts on the first joint Sino-Russian Arctic scientific expedition.[128] Lasting 32 days, the expedition took place aboard the Russian research vessel Academic M.A. Lavrentyev in the Chukchi Sea and the East Siberian Sea within the Russian EEZ, previously inaccessible to China. The researchers conducted comprehensive research in the fields of marine geology, physical oceanography, marine chemistry, and atmospheric chemistry. Shi Xuefa, the lead Chinese scientist on the expedition, emphasised that the research conducted would make China's understanding of the Arctic marine environment "more complete, comprehensive, and systematic."[129] This was followed by a second joint scientific expedition to the region in 2018, where Chinese scientist ventured to the Russian Arctic waters (including the Chukchi, East Siberian, and Laptev seas) to conduct research in marine geology, hydrometeorology, and biology.[130] Chinese reports highlighted the expedition's contribution to the development of the PSR, as it focused on key areas along the NSR and acquired data related to Arctic navigation, including fog, snow, and cloud cover.[131] Despite the challenges posed by the COVID-19 pandemic, Chinese and Russian scientists managed to conduct an expedition in 2020. Having jointly developed an implementation plan, the Russian scientists conducted on-site research while the Chinese scientists participated remotely due to travel restrictions, with the Russians sharing the collected data with them.[132]

At the institutional level, Chinese and Russian science organisations are strengthening their ties through the establishment of joint research centres. In 2019, building on their previous positive interactions, the Russian Academy of Sciences and the Qingdao National Laboratory for Marine Science and Technology agreed to establish the Chinese-Russian Arctic Research Centre. The centre focuses on studying the climate, geology, and ecology of the Arctic to forecast ice conditions on Arctic shipping routes, and provide recommendations for Arctic economic development.[133] Similarly, HEU and St. Petersburg State Marine Technical University established the China-Russia Polar Technology and Equipment "Belt and Road" joint laboratory, focusing on Arctic shipping, energy, and science cooperation.[134] The HIT established the Sino-Russian Polar Engineering and Research Centre, which aims to design ice-resistant platforms for Arctic use, study the behaviour of concrete structures in polar zones, analyse ice impacts on ships, and facilitate student and researcher exchanges.[135]

In terms of academic exchanges, Chinese and Russian specialists convene annually at the China-Russia Arctic Forum (中俄北极论坛) co-organised by the Ocean University of China and St. Petersburg State University since 2012. During these meetings, academics from various Chinese and Russian institutions present their views on a wide range of issues, such as China's participation in Arctic resources development, navigation on the NSR, and the impact of geopolitical tensions on regional cooperation.[136] In addition to this forum, other Chinese higher education institutions engage their Russian partners in discussions related to the Arctic. For instance, in 2021, Jilin University organised a forum in cooperation with Russian partners to explore Sino-Russian cooperation in the sustainable development of the Arctic.[137]

The extensive Russian Arctic territory, along with Russian experience and capabilities, can also assist the PRC's armed forces in improving certain capabilities, including cold-weather training and naval drills. While substantial engagement to advance Arctic military cooperation has not yet taken place, the deepening of bilateral relations and the ever-expanding operational radius of the PLAN indicate the potential for such cooperation. Given that China has already conducted exercises with Russian forces in the Baltic and Bering seas, the possibility of further military collaboration extending into the Arctic should not be dismissed.[138] Additionally, opportunities exist in the field of missile defence. Deng argues that in such field of Arctic strategic security, Russia and China share common interests and concerns and should strengthen what he labels as "back-to-back strategic support." He suggests that the Arctic is where both countries face a shared strategic threat, particularly the US anti-missile system, providing a basis for strategic security cooperation. Deng further notes that as Russia is surrounded by the US alliance system on three sides (Europe, the Arctic, and the Pacific), China could serve as a strategic rear for Russia in addressing its Arctic security issues.[139] The notion of a strategic rear, however, can be reversed and applied to Russia in light of China's security environment. In this regard, as noted by Hsiung, the PRC would benefit greatly from early warning assistance Russia could provide regarding American intercontinental ballistic missiles (ICBMs) flying over the Arctic region,[140] thereby contributing to the security of the Chinese northern flank.

It appears that the Sino-Russian Arctic partnership can continue to evolve, despite Russia facing growing political and military pressures from Western states due to its actions in Ukraine. Chinese capital and technology played an important role in the development of Russian Arctic energy projects following Russia's annexation of Crimea in 2014. Despite challenges in their Arctic relations, including the seemingly low rate of implementation of some Arctic projects,[141] and claims that they have diverging regional agendas,[142] the mutual engagement between China and Russia in the Arctic has persisted. The Sino-Russian partnership was reaffirmed in February 2022 when both states declared their willingness "to continue consistently intensifying practical cooperation for the sustainable development of the Arctic."[143] While it is true that some Chinese SOEs temporarily halted their commercial activities in Russia to avoid entanglement in US and EU sanctions in response to the Russian invasion of Ukraine in 2022, Eiterjord argues that this cautious approach may be short-lived, as unlocking Russia's energy potential is key to the development of China's Polar Silk Road.[144] Chinese investors in Russian Arctic LNG projects, CNOOC and CNPC, remain committed to the projects, and both Putin and Xi have called for closer bilateral energy cooperation.[145] Following their meeting in March 2023, reports indicated that the Russians were open to giving the Chinese a more significant role in the development of the resource and shipping potential of the Russian Arctic.[146] As Russian strategic options diminish and their concerns about Western threats increase, China finds itself in an advantageous position to choose whether to further involve itself in the Russian Arctic and under what conditions.

The Arctic in Sino-American Relations

Similar to the Russian Arctic zone, the North American Arctic encompasses vast and sparsely populated territories of the US (Alaska) and Canada (Yukon, the Northwest Territories, Nunavut, and the northern parts of Quebec). The region is rich in living and nonliving resources, including large quantities of oil, natural gas, and minerals like diamonds, iron ore, and gold. Mining and other resource extraction industries form the backbone of commercial activity in the North American Arctic.[147] Additionally, the shipping potential of the Northwest Passage, which runs through the Beaufort Sea and the Canadian Archipelago, adds to the economic attraction of the region. This passage connects the Pacific and Atlantic Oceans and, like the NSR, can significantly reduce travel time and distance between Northeast Asia and the North Atlantic region. During the Cold War, the Arctic region held strategic significance for the United States as a military arena for its war games with the Soviet Union. To guard against potential Soviet air and missile strikes, the US established a continental-wide deterrence and defence-radar network known as the Distant Early Warning Line, stretching from Alaska to Greenland. However, with the collapse of the Soviet Union and diminished threat perceptions, the Arctic lost some of its military utility. Consequently, in the post–Cold War era, American and Canadian interests in the Arctic waned, with the exceptions of American global missile defence architecture based in Alaska and the focus on Arctic scientific research.[148]

China's engagement with the North American Arctic throughout much of the post–Cold War era remained somewhat limited yet generally conducted in a cooperative manner. In the 1990s, some of China's first scientific expeditions to the Arctic took place in Alaska. The US also played a mediating role in facilitating China's accession to the AC, reportedly brokering an agreement acceptable to other AC members regarding the acceptance of new observer states (some AC members, including Canada, were not enthusiastic about inviting new observers to the AC).[149] American and Chinese Arctic specialists also participated in various dialogues, enabling the exchange of views on Arctic interests and strategies. For instance, in 2017, Tongji University and the Centre for Strategic and International Studies organised the third Sino-American Arctic Social Science Symposium, where experts and officials from both nations discussed topics such as Arctic governance, northern geopolitics, and sustainable development.[150] China also expressed interest in the region's shipping potential when the Chinese Maritime Safety Administration published guidelines for shipping on the NWP in 2016. On that occasion, Liu Pengfei, spokesperson for China's transportation ministry, noted that "once this route is commonly used, it will directly change global maritime transportation and have a profound influence on international trade, the world economy, capital flow, and resource exploitation," and that "there will be ships with Chinese flags sailing through this route in the future."[151] China's potential commercial use of the NWP was again highlighted in 2017 when its research vessel, the Xuelong, made a historic passage through the NWP. Chinese state-run media announced that the voyage had assisted in accumulating a wealth of navigational experience for future Chinese ships.[152] There was also much discussion the same year about Chinese corporations funding the development of Alaska's LNG potential.[153] However, this project never materialised. Instead, it became a visual representation of the breakdown in bilateral relations that soon followed, and which effectively halted any meaningful engagement between the US and China in the Arctic.

Currently, China and the US are engaged in a strategic competition that reverberates throughout the international system, and across regions, including the Arctic. The 2022 US National Security Strategy portrays the PRC as the only world power that has the economic, political, military, and technological capabilities to reshape the international order, using its influence over international organisations "to create more permissive conditions for its own authoritarian model."[154] Rep. Mike Gallagher, Chairman of the United States House Select Committee on Strategic Competition between the United States and the Chinese Communist Party, characterises the strategic competition between the US and the PRC as "an existential struggle over what life will look like in the 21st century."[155] On the other side of the Pacific, the CCP seems to be convinced that the US is abusing its power to hinder the PRC's further development and, sticking to the so-called Cold War mentality, creating political blocs to stoke confrontation and suppress China's engagement with countries worldwide. In response to American pressures, the CCP has launched a global diplomatic campaign, offering its own solutions to the perceived shortcoming of a US-dominated world and warning about the detrimental effects of American hegemony.[156] This strategic competition is not transitory but is expected to persist for the foreseeable future, as both countries acknowledge this new reality.[157]

In response to perceived advances by Russia and China in the Arctic, the US government has been increasing its strategic presence in the region. This includes upgrading its military installations on Greenland, redeploying its air force to Iceland, and constructing new facilities in Northern Norway capable of accommodating US armed forces.[158] The US, along with its allies, has conducted joint military exercises above the Arctic Circle and naval patrols in the Barents Sea.[159] Additionally, after 40 years, the decision was made to upgrade its ageing fleet of heavy icebreakers, with new vessels expected to be operational in the late 2020s.[160] At the policy level, the Trump administration criticised Chinese regional intentions, while the Biden administration, in its 2022 National Strategy for the Arctic Region, acknowledges Chinese ambitions to increase its influence in the Arctic.[161]

Chinese analysts view this American strategic re-engagement with the region with a degree of scepticism. The increased military activity by Russia in the Arctic is not seen as problematic from the Chinese perspective. Yang notes that China does not object to the Russian military being deployed in its own Arctic zone. He points out that this is an expression of Russia's right to protect its national interests and enhance its governance capabilities, particularly in Arctic search and rescue operations.[162] On the other hand, some Chinese commentators believe that the course of action taken by the US and its allies contributes to regional tensions. For example, Li Yun from the PLA's Academy of Military Sciences notes that the US, through the expansion of its Arctic military infrastructure, is not only strengthening its own military power in the region but also fuelling a revival of the Cold War mentality of bloc confrontation.[163] At the policy level, a commentary in the *PLA Daily* suggests that the American National Strategy for the Arctic Region indicates a desire to preserve US leadership in the region (that is its Arctic hegemony), potentially leading to the establishment of discriminatory standards that exclude some countries from participating in Arctic affairs.[164] From Xiao Yang's perspective, the US is pursuing a strategy of resistance vis-à-vis the PRC in the Arctic. According to Xiao, when China introduced the PSR, the US realised that it will be unable to stop Chinese advances in the Arctic. Therefore the US is implementing a series of political, military, and economic measures (such as a larger policy focus on the Arctic in American foreign policy discourses, new funding for Coast Guard vessels and military installations in Alaska) all in the hopes of slowing the decline of its hegemony, preserving its regional leadership, and competing with the Chinese PSR.[165] Others see the inclusion of the PRC in American Arctic strategic documents and debates as a new version of the so-called China threat theory – the China Arctic Threat Theory (中国北极威胁论). In a recent analysis of this theory, Liu Dan notes that the US believes the PRC has military plans for the Arctic and, together with Russia, has altered the regional balance of power. Consequently, in order to rationalise its own military presence in the region and confront China and Russia, the US exaggerates and compares the Arctic to the South China Sea, leading to heightened regional threat perceptions.[166] US actions are sometimes contrasted with China's agendas in the Arctic which, according to Chinese discourses, are supposedly focused on regional partnerships and not the creation of exclusive blocs or cliques.[167]

Regarding the Canadian Arctic, the economic and scientific activities of the PRC remain limited due to deteriorating bilateral relations between China and Canada, exacerbated by the arrests of Huawei's Meng Wanzhou and the Canadians Michael Kovrig and Michael Spavor in 2018. While there is potential for the development of natural resources in the Canadian Arctic, and Chinese companies have shown interest by providing funding for some projects, Chinese SOEs face challenges in establishing themselves in the Canadian mining sector due to increased scrutiny on national security grounds. For instance, in 2020, Shandong Gold, a provincial-level Chinese SOE was prevented from acquiring a gold mine in the Canadian Arctic due to security concerns.[168] Another potential source of friction between Canada and the PRC could be the status of the NWP. Canada considers it as internal waters, while other countries, notably the US, view it as an international strait. Although the PRC is interested in the shipping potential of Arctic sea routes, it has not taken a decisive stance on the status of the NWP. It did, however, notice that "Canada has imposed some restrictions on the use of the Northwest Passage, asking foreign vessels to inform the Canadian side and get permission before entering or crossing its exclusive economic zone and territorial waters."[169] This was followed in 2017 by the passage of the Chinese research icebreaker Xuelong through the NWP during which the Chinese adhered to Canadian rules and regulations.

The Arctic in Sino-Nordic Relations

The Nordic Arctic, comprising the northern parts, regions, and counties of Finland, Sweden, Norway, Denmark, and Iceland, differs in certain aspects from the rest of the Arctic region. This is due to a higher concentration of wealth and state presence, both nationally and internationally, particularly in Scandinavia.[170] The Nordic Arctic boasts stable and advanced industries, including resource extraction such as oil, gas, mineral ores, and boreal forests. It also has well-developed regional transportation infrastructure and a high concentration of research institutions and facilities such as UiT the Arctic University of Norway, Nord University, and the University of Lapland.[171] Internationally, the Nordic Arctic holds great importance on the foreign policy agendas of the Nordic states, with the High North being Norway's primary foreign policy priority. Moreover, this region hosts significant gatherings focused on Arctic affairs, like the Arctic Circle Assembly in Iceland, and serves as the location of the AC secretariat and the Indigenous Peoples' Secretariat in Tromsø, Northern Norway. These factors make the Nordic Arctic a potentially attractive area for China's engagement with the Arctic.

For some time, this appeared to be the case. Even though the PRC "froze" its relations with Norway from 2010 to 2016 following the awarding of the Nobel Peace Prize to Chinese dissident Liu Xiaobo, this did not prevent China from deepening its ties with the rest of the Nordic states. In 2012, Minister of Land and Resources Xu Shaoshi and SOA head Liu Cigui visited Greenland to discuss cooperation in resource extraction, Arctic research, and glacier observations.[172] The same year, Chinese Premier Wen Jiabao visited Iceland, resulting in the signing of several agreements, including a framework agreement on Arctic cooperation and an MoU on marine and polar science and technology cooperation.[173] Shortly after, the PRC and Iceland signed a Free Trade Agreement (FTA), the first of its kind between China and a European country.[174] Additionally, in 2017, China and Finland decided to establish a future-oriented new-type cooperative partnership by "enhancing political mutual trust and expanding and deepening cooperation to the benefit of the two peoples."[175] Through this partnership, the two countries also pledged to increase economic and technological cooperation in Arctic marine industry, Arctic geology, Arctic-related research, maritime safety, and tourism.[176]

The Nordic Arctic has also served as an important location for China's terrestrial research presence in the region. China's research stations are situated on Norwegian and Icelandic territory, and discussions have taken place regarding the establishment of a Chinese research station in Greenland. There have also been reports of the Chinese PRIC's interest in renting and expanding Finland's Kemijärvi airport in eastern Lapland as a base for conducting climate and environmental research flights over the Central Arctic Ocean.[177] Deng suggests that cooperation with Nordic countries presents an opportunity for China to engage in areas of "low politics," including maritime safety, search and rescue, emergency responses, and data sharing. He also notes that the Nordics are global leaders in various polar-related technological fields, such as ice-class vessel construction, unmanned systems, automation, and smart technologies. Deepening cooperation in these areas could, according to Deng, enhance China's Arctic capabilities and situational awareness.[178]

Opportunities have also emerged in the commercial sector, particularly in relation to the mining potential in Greenland, where several projects currently involve Chinese backing. While the existing level of Chinese engagement in the Greenlandic mineral sector appears to be limited, some international observers suggest that China may be adopting an active wait-and-see approach, monitoring developments on the island for future opportunities.[179] In Iceland, Chinese company Sinopec is partnering with local geothermal companies to develop solutions and technological expertise that could be applied to geothermal projects on the Chinese mainland. There also was enthusiasm, particularly on the Nordic side, regarding the inclusion of the Arctic within the framework of the BRI and the potential it could bring to the region. One example of this enthusiasm was the planned Arctic Corridor project, which aimed to establish the Norwegian town of Kirkenes as a transhipment hub and then connect it to the rest of the European mainland via a railway cutting across Finland. China was viewed as a potential user and partial investor in this project.[180]

However, the PRC's engagement with the Nordic Arctic is now facing its own set of challenges in the increasingly restrictive Arctic security environment. Some Nordic states have become more sensitive to China's presence in the region and the security implications associated with it. In 2016, the Danish government blocked the sale of the decommissioned Grønnedal military base in southern Greenland to the Chinese mining company General Nice. Security concerns were cited as the main reason for this decision, as Denmark did not want to jeopardise its close relations with the US by granting further access to Chinese actors on the island.[181] Later, in 2019, after what appeared to be direct intervention from the US, the Danish government announced that a consortium of Danish financial institutions would provide funding for the development of three airports in Greenland, side-lining the China Communications Construction Company, a powerful Chinese SOE, which had been bidding for the airport construction project.[182] The American strategic refocus on the Arctic could further complicate Chinese involvement in Greenland, which some academics in the PRC consider to be "the true strategic fulcrum of the Polar Silk Road."[183]

Similarly, Norwegian security agencies view the PRC as "a significant intelligence threat against Norwegian interests."[184] Norway's domestic intelligence and security service, the Police Security Service, stated in its 2023 National Threat Assessment report that the PRC may seek to enhance its presence in the High North by purchasing strategically located properties that could facilitate intelligence operations and create economic dependencies vulnerable to future exploitation, posing challenges to Norwegian security.[185] With regards to Finland, Puranen and Kopra note that planned Chinese economic and scientific projects did not materialise, primarily due to security reasons. They add that optimism about China that prevailed in Finland in the late 2010s has been replaced by more scepticism and suspicion regarding its Arctic intentions.[186]

Given the current state of Arctic affairs and the changing perceptions of the PRC in the region, it is reasonable to conclude that the best opportunities for the PRC to advance its interests in the region lie in a partnership with Russia. The Russian leadership appears to be re-evaluating its strategic options and becoming more accepting of China's involvement in its Arctic zone. However, in the North American Arctic, Chinese options will remain limited as all forms of engagement will be carefully scrutinised under the guise of national security. In the Nordic Arctic, views about China and its Arctic involvement have soured considerably. Nevertheless, as Europe explores the idea of its strategic autonomy, which China seems to enthusiastically support, there is still the possibility that some actors in the Nordic Arctic may be open to cooperation with the PRC in various forms.

Notes

1 For example, see the speech by Chinese Premier Li Qiang at the Boao Forum for Asia Annual Conference in 2023: Li Qiang, "Following the Vision of a Community with a Shared Future for Mankind and Bringing More Certainty to World Peace and Development," *Ministry of Foreign Affairs of the PRC*, March 30, 2023, www.fmprc.gov.cn/eng/wjdt_665385/zyjh_665391/202303/t20230331_11052581.html.
2 Katherine Morton, "China's Global Governance Interactions," in *China and the World*, ed. David Shambaugh (New York: Oxford University Press, 2020), 175–176.
3 "Full Text: China's Arctic Policy," *Xinhua*, January 26, 2018, www.xinhuanet.com/english/2018-01/26/c_136926498.htm.
4 Oran R. Young, "Is It Time for a Reset in Arctic Governance?" *Sustainability* 11, no. 16, 4497, (2019): 1.
5 Young, "Is It Time for a Reset in Arctic Governance?" 3.
6 Oran R. Young, "Governing the Arctic: From Cold War Theatre to Mosaic of Cooperation," *Global Governance* 11 (2005): 10.
7 Olav Schram Stokke, "Environmental Security in the Arctic: The Case for Multilevel Governance," *International Journal* 66, no. 4 (2011): 835–848.

8   Young, "Is It Time for a Reset in Arctic Governance?" 3.
9   "About the Arctic Council," *Arctic Council*, n.d., https://arctic-council.org/about/.
10 Michael R. Pompeo, "Looking North: Sharpening America's Arctic Focus," *US State Department*, May 6, 2019, https://2017-2021.state.gov/looking-north-sharpening-americas-arctic-focus/index.html.
11 Young, "Is It Time for a Reset in Arctic Governance?" 3.
12 For a nuanced discussion about the establishment of the AC, see Svein Vigeland Rottem, *The Arctic Council: Between Environmental Protection and Geopolitics* (Singapore: Palgrave Macmillan, 2020), 1–17.
13 The full list of the Permanent Participant includes the Aleut International Association, the Arctic Athabaskan Council, the Gwich'in Council International, the Inuit Circumpolar Council, the Russian Association of Indigenous Peoples of the North, and the Saami Council, "About the Arctic Council," *Arctic Council*, n.d., https://arctic-council.org/about/.
14 AC WGs include the Arctic Contaminants Action Program, Arctic Monitoring and Assessment Programme, Conservation of Arctic Flora and Fauna, Emergency Prevention, Preparedness and Response, Protection of the Arctic Marine Environment and the Sustainable Development Working Group, "Working Groups," *Arctic Council*, n.d., https://arctic-council.org/about/working-groups/. The Expert Group is on Black Carbon and Methane, "Expert Group in Support of Implementation of the Framework for Action on Black Carbon and Methane (EGBCM)," *Arctic Council*, n.d., https://arctic-council.org/projects/expert-group-in-support-of-implementation-of-the-framework-for-action-on-black-carbon-and-methane-egbcm/.
15 For a detailed overview of the structure and procedures of the AC see Rottem, *The Arctic Council: Between Environmental Protection and Geopolitics*, 19–45.
16 "Arctic Council Observers," *Arctic Council*, n.d., https://arctic-council.org/about/observers/.
17 Arctic Council, *Declaration on the Establishment of the Arctic Council* (Ottawa, 1996), https://oaarchive.arctic-council.org/bitstream/handle/11374/85/EDOCS-1752-v2-ACMMCA00_Ottawa_1996_Founding_Declaration.PDF?sequence=5&isAllowed=y.
18 Lassi Heininen, "Special Features of Arctic Geopolitics – A Potential Asset for World Politics," in *The GlobalArctic Handbook*, eds. Matthias Finger and Lassi Heininen (Cham, Switzerland: Springer International Publishing, 2019), 215–234.
19 Juha Käpylä and Harri Mikkola, "Contemporary Arctic Meets World Politics: Rethinking Arctic Exceptionalism in the Age of Uncertainty," in *The GlobalArctic Handbook*, eds. Matthias Finger and Lassi Heininen (Cham, Switzerland: Springer International Publishing, 2019), 153.
20 Lassi Heininen, Heather Exner-Pirot and Joël Plouffe, "Governance & Governing in the Arctic: An Introduction to Arctic Yearbook 2015," in *Arctic Yearbook 2015*, eds. Lassi Heininen, Heather Exner-Pirot and Joël Plouffe (Akureyri, Iceland: Northern Research Forum, 2015), https://arcticyearbook.com/arctic-yearbook/2015/12-yearbook/2015-arctic-governance-and-governing/121-governance-governance-in-the-arctic-an-introduction-to-arctic-yearbook-2015.
21 For an extensive discussion on Arctic Exceptionalisms see P. Whitney Lackenbauer and Ryan Dean, "Arctic Exceptionalisms," in *The Arctic and World Order*, eds. Kristina Spohr, Daniel S. Hamilton and Jason C. Moyer (Washington, DC: Johns Hopkins University, 2020), 327–355. For a contradictory view, that the Arctic is not exceptional but that it was always a part of the broader international system, see Rasmus Gjedssø Bertelsen, "Arctic Security in International Security," in *Routledge Handbook of Arctic Security*, eds. Gunhild Hoogensen Gjørv, Marc Lanteigne and Horatio Sam-Aggrey (London: Routledge, 2020): 57–68.
22 Young, "Is It Time for a Reset in Arctic Governance?" 4–7.
23 Kristina Spohr and Daniel S. Hamilton, "From Last Frontier to First Frontier: The Arctic and World Order," in *The Arctic and World Order*, eds. Kristina Spohr,

Daniel S. Hamilton and Jason C. Moyer (Washington, DC: Johns Hopkins University, 2020), 11–12.

24 Eric Roston, "How a Melting Arctic Changes Everything, Part 3: The Economic Arctic," *Bloomberg*, December 29, 2017, www.bloomberg.com/graphics/2017-arctic/the-economic-arctic/.

25 Oran R. Young, "Constructing the 'New' Arctic: The Future of the Circumpolar North in a Changing Global Order," *Outlines of Global Transformations: Politics, Economics, Law* 12, no. 5 (2019): 13.

26 Marc Lanteigne, "By Any Other Name? Defining an Arctic Stakeholder Is Becoming More Complex," *Over the Circle*, October 7, 2019, https://overthecircle.com/2019/10/07/by-any-other-name-defining-an-arctic-stakeholder-is-becoming-more-complex/.

27 Beate Steinveg, "The Role of Conferences Within Arctic Governance," *Polar Geography* 44, no. 1 (2021): 37–54.

28 Julie Wilhelmsen and Kristian Lundby Gjerde, "Norway and Russia in the Arctic: New Cold War Contamination?" *Arctic Review on Law and Politics* 9 (2018): 382–407.

29 Käpylä and Mikkola, "Contemporary Arctic Meets World Politics: Rethinking Arctic Exceptionalism in the Age of Uncertainty," 157.

30 Hilde-Gunn Bye, "Leaving Its Arctic Reluctance Behind: The Re-Emergence of US Security Policy Focus Towards the European High North and Its Implications for Norway," *The Polar Journal* 10, no. 1 (2020): 82–101.

31 James Kenneth Wither, "An Arctic Security Dilemma: Assessing and Mitigating the Risk of Unintended Armed Conflict in the High North," *European Security* 30, no. 4 (2021): 649–666.

32 "Joint Statement on Arctic Council Cooperation Following Russia's Invasion of Ukraine," *US Department of State*, March 3, 2022, www.state.gov/joint-statement-on-arctic-council-cooperation-following-russias-invasion-of-ukraine/.

33 The White House, *National Strategy for the Arctic Region* (Washington, DC, 2022), www.whitehouse.gov/wp-content/uploads/2022/10/National-Strategy-for-the-Arctic-Region.pdf.

34 "Full Text: China's Arctic Policy," *Xinhua*.

35 Yang Jian, 北极治理新论 [*New Perspectives on the Arctic Governance*] (Beijing: 时事出版社 [Current Affairs Press], 2014), 297–298.

36 Marc Lanteigne, *China's Emerging Arctic Strategies: Economics and Institutions* (Reykjavik: University of Iceland Press, 2014), 36.

37 Jia Xiaopan, "People's Republic of China – Observer Report 2020," *Arctic Council*, November 30, 2020, https://oaarchive.arctic-council.org/handle/11374/2717.

38 For example, see Li Linlin, "The People's Republic of China 2018 Observer Review Report," *Arctic Council*, May 31, 2018, https://oaarchive.arctic-council.org/handle/11374/2251.

39 Chih Yuan Woon and Klaus Dodds, "The Fear of Bypass: The Flexible Diplomacy of Jacking-Up and Stretching Out Singapore's Arctic Connections," in *'Observing' the Arctic: Asia in the Arctic Council and Beyond*, eds. Chih Yuan Woon and Klaus Dodds (Cheltenham, UK and Northampton: Edward Elgar Publishing, 2020), 215–216.

40 Adam Stepien, "Incentives, Practices and Opportunities for Arctic External Actors' Engagement with Indigenous Peoples: China and the European Union," in *Arctic Law and Governance: The Role of China, Finland and the EU*, eds. Timo Koivurova, Tianbao Qin, Sébastien Duyck and Tapio Nykänen (Oxford and Portland: Hart Publishing, 2017), 217.

41 Yang Xiaoning, "China's 2016 Observer Activities Report," *Arctic Council*, November 25, 2016, https://oaarchive.arctic-council.org/handle/11374/1860.

42 Mia M. Bennett, "Scale-Jumping in the Arctic Council: Indigenous Permanent Participants and Asian Observer States," in *'Observing' the Arctic: Asia in the Arctic Council and Beyond*, eds. Chih Yuan Woon and Klaus Dodds (Cheltenham, UK and Northampton: Edward Elgar Publishing, 2020), 72.

43 Peng Qiuhong and Lu Junyuan, "原住民权利与中国北极地缘经济参与" [The Rights of Indigenous Peoples and China's Participation in Arctic Geo-economy], 世界地理研究 [*World Regional Studies*] 22, no. 1 (2013): 32–38.

44 "Full Text: China's Arctic Policy," *Xinhua*.

45 Gao Feng – Special Representative for Climate Change Negotiations of the Foreign Ministry of China, *Speech at the 2016 Arctic Circle Assembly*, Iceland (Video), https://vimeo.com/189527301.

46 Zhao Pengfei, "中国坦荡参与北极治理" [China Openly Participates in Arctic Governance], 人民日报 [*People's Daily*], January 30, 2018, http://paper.people.com.cn/rmrbhwb/html/2018-01/30/content_1833530.htm.

47 Wang Libin, "我国将为北极可持续发展贡献中国智慧" [My Country Will Contribute Chinese Wisdom to the Sustainable Development of the Arctic], *Xinhua*, May 10, 2019, https://web.archive.org/web/20211028085644/www.xinhuanet.com/politics/2019-05/10/c_1124478690.htm.

48 Chih Yuan Woon, "Framing the 'Polar Silk Road' (冰上丝绸之路): Critical Geopolitics, Chinese Scholars and the (Re)Positionings of China's Arctic Interests," *Political Geography* 78 (2020): 1–2.

49 "外交部：中国在北极事务中发挥作用不越位、不缺位" [Ministry of Foreign Affairs: China's Role in Arctic Affairs Is to Not Overstep, Not Being Absent], 中国新闻网 [*China News*], January 26, 2018, www.chinanews.com.cn/gn/2018/01-26/8433829.shtml.

50 "Full Text: China's Arctic Policy," *Xinhua*.

51 "Keynote Speech by Vice Foreign Minister Zhang Ming at the China Country Session of the Third Arctic Circle Assembly," *Ministry of Foreign Affairs of the PRC*, October 17, 2015, www.fmprc.gov.cn/eng/wjdt_665385/zyjh_665391/201510/t20151017_678393.html.

52 Marc Lanteigne, "Considering the Arctic as a Security Region: The Roles of China and Russia," in *Routledge Handbook of Arctic Security*, eds. Gunhild Hoogensen Gjørv, Marc Lanteigne and Horatio Sam-Aggrey (London: Routledge, 2020), 317.

53 Xu Hong, "北极治理与中国的参与" [Arctic Governance and China's Participation], 边界与海洋研究 [*Journal of Boundary and Ocean Studies*] 2, no. 2 (2017): 6.

54 Xu, "北极治理与中国的参与" [Arctic Governance and China's Participation], 6.

55 Gao Feng, "China and the Arctic: Engaging After the Global Pandemic," *Arctic Circle Assembly 2022*, Youtube video, www.youtube.com/watch?v=qTtSOpKJ55A&ab_channel=ArcticCircle.

56 Zhang Yao, "Ice Silk Road Framework Welcomed by Countries, Sets New Direction for Arctic Cooperation," *Global Times*, April 11, 2019, https://brgg.fudan.edu.cn/en/articleinfo_395.html.

57 Sun Kai and Wu Junhuan, "北极治理新态势与中国的深度参与战略" [New Developments in Arctic Governance and China's Deepening Involvement Strategy], 国际展望 [*Global Review*], no. 6 (2015): 71–74.

58 Luo Huijun and Zha Yunlong, "北极治理'内卷化'与中国应对" [The Involution of the Arctic Governance and China], 中国海洋大学学报 [*Journal of Ocean University of China*], no. 4 (2021): 82–84.

59 For a discussion regarding the CAO Fisheries agreement, see: Alexander N. Vylegzhanin, Oran R. Young and Paul Arthur Berkman, "The Central Arctic Ocean Fisheries Agreement as an Element in the Evolving Arctic Ocean Governance Complex," *Marine Policy* 118 (2020): 1–10.

60 Deng Beixi, 北极安全研究 [*Arctic Security Studies*] (Beijing: 海洋出版社 [Ocean Press], 2020), 203–206.

61 Zhao Hua and Kuang Zengjun, "中国学者的北极问题研究---基于中国国际政治类核心杂志 (2007–2016)" [Review of the Arctic Study in China Based on Articles in Chinese Core Journals of International Politics], 战略决策研究 [*Journal of Strategy and Decision-Making*] 4 (2017): 86.

62 "Arctic Council Observer Manual for Subsidiary Bodies," *Arctic Council*, October 4, 2016, https://oaarchive.arctic-council.org/handle/11374/939.

63 Guo Peiqing, "An Analysis of New Criteria for Permanent Observer Status on the Arctic Council and the Road of Non-Arctic States to Arctic," *KMI International Journal of Maritime Affairs and Fisheries* 4, no. 2 (2012): 31.

64 Xiao Yang, "北极理事会'域内自理化'与中国参与北极事务路径探析" [On the 'Field Self-Governance' of the Arctic Council and the Path for China to Participate in the Arctic Affairs], 现代国际关系 [*Contemporary International Relations*] 1 (2014): 52–54.

65 Pan Min, "论中国参与北极事务的有利因素, 存在障碍及应对政策," [Study on Favorable Factors, Barriers and Coping Strategies of China's Participation in Arctic Affairs], 中国软科学 [*China Soft Science*] 6 (2013): 16.

66 Ni Haining and Li Ming, "北极军事化趋势堪忧" [Arctic Militarization Is Worrisome], 解放军报 [*PLA Daily*], February 12, 2016, https://web.archive.org/web/20160213092002/https://mod.gov.cn/opinion/2016-02/12/content_4640521.htm.

67 Li Zhixin, "北极地区军事化愈演愈烈" [Arctic Militarization Intensifies], 中国青年报 [*China Youth Daily*], December 6, 2018, http://m.xinhuanet.com/mil/2018-12/06/c_1210009928.htm.

68 Li Yuan, "美新版北极战略加剧极地博弈" [The New US Arctic Strategy Intensifies Polar Games], 中国国防报 [*China National Defence Daily*], October 14, 2022, http://military.people.com.cn/n1/2022/1014/c1011-32545084.html.

69 "冷战思维死灰复燃 北极地区军事博弈愈演愈烈" [The Cold War Mentality Is Revived, Arctic Military Games Intensify], 解放军报 [*PLA Daily*], June 9, 2022, www.news.cn/mil/2022-06/09/c_1211655269.htm.

70 Jiang Yin'an, "北极安全形势透析：动因，趋向与中国应对" [The Situation of the Arctic Security: Causes, Trends and China's Approach], 边界与海洋研究 [*Journal of Boundary and Ocean Studies*] 5, no. 6 (2020): 105.

71 Zuo Pengfei, 极地战略问题研究 [*A Study on Polar Strategy*] (Beijing: 时事出版社 Current Affairs Press, 2018), 47.

72 Chinese scholars are, of course, not the only ones proposing changes to the current Arctic governance system, as international scholars have done so as well. For example, see Oran R. Young, "Is It Time for a Reset in Arctic Governance?" *Sustainability* 11, no. 16, 4497, (2019): 1–12.

73 Yang Jian, *Arctic Governance and China's Engagement* (Shanghai: Shanghai Institutes of International Studies, 2022), 22.

74 Pan Min and Xu Liling, "超越门罗主义：北极科学部长级会议与北极治理机制革新" [Beyond Monroe Doctrine: The Arctic Science Ministerial and the Innovation of Arctic Governance Regime], 太平洋学报 [*Pacific Journal*] 29, no. 1 (2021): 92–100.

75 Li Zhenfu and Liu Tongchao, "北极航线地缘安全格局演变研究" [Research on the Evolution of the Arctic Route Geo-Security Structure], 国际安全研究 [*Journal of International Security Studies*] 33, no. 6 (2015): 81–105.

76 Ding Huang and Zhu Baolin, "基于'命运共同体'理念的北极治理机制创新" [The Innovation of the Arctic Governance Mechanism Based on the Concept of the "Community of Destiny"], 探索与争鸣 [*Exploration and Free Views*], no. 3 (2016): 94–99.

77 Xi Jinping, "Work Together to Build a Community of Shared Future for Mankind," *Xinhua*, January 19, 2017, www.xinhuanet.com/english/2017-01/19/c_135994707.htm.

78 Bai Jiayu, "中国积极参与北极公域治理的路径与方法 – 基于人类命运共同体理念的思考" [The Path and Means of China's Active Participation in Arctic Public Governance – Based on the Vision of a Community with a Shared Future for Mankind], 学术前沿 [*Frontiers*], no. 23 (2019), www.rmlt.com.cn/2020/0721/587499.shtml.

79 Li Zhenfu et al. "冰上丝绸之路与北极命运共同体构建研究" [Research on the Construction of the Polar Silk Road and Arctic Destiny Community], 社会科学前沿 [*Advances in Social Sciences*] 8, no. 8 (2019): 1420–1421.

80 Xu Guangmiao, "变动世界中的北极秩序：生成机制与变迁逻辑" [The Arctic Order in a Changing World: Generative Mechanism and Transition Logic], 俄罗斯东欧中亚研究 [*Russian Central Asian and East European Studies*], no. 1 (2021): 123.

81 Guo Peiqing, "对北极治理机制的新探索和新突破" [New Explorations and New Breakthroughs in the Arctic Governance Mechanism], 中华读书报 [*China Reading Weekly*], September 29, 2021, https://epaper.gmw.cn/zhdsb/html/2021-09/29/nw.D110000zhdsb_20210929_1-19.htm.

82 Henry Tillman, Yang Jian and Egill Thor Nielsson, "The Polar Silk Road: China's New Frontier of International Cooperation," *China Quarterly of International Strategic Studies* 4, no. 3 (2018): 347–348.

83 "Full Text: China's Arctic Policy," *Xinhua*.

84 Sun Kai, "从愿景到行动：推进"冰上丝绸之路"建设正当其时" [From Vision to Action: The Right Time for Advancing the Development of the Polar Silk Road], 中国社会科学网 [*Chinese Social Sciences Net*], February 8, 2018, https://web.archive.org/web/20210717231948/www.cssn.cn/zzx/gjzzx_zzx/201802/t20180208_3845186.shtml.

85 Marc Lanteigne, "Only Connect? The Polar Silk Road and China's Geoeconomic Policies," in *Nordic-Baltic Connectivity with Asia via the Arctic: Assessing Opportunities and Risks*, eds. Bart Gaens, Frank Jüris and Kristi Raik (Tallinn, Estonia: International Centre for Defence and Security, 2021): 115.

86 Yang, *Arctic Governance and China's Engagement*, 123.

87 Dong Yongzai, "极地安全：国家安全的新疆域" [Polar Security: A New Frontier for National Security], 光明日报 [*Guangming Daily*], April 25, 2021, https://news.gmw.cn/2021-04/25/content_34790839.htm.

88 For example, see Zhang Jiajia, "中国北极话语权及其提升路径研究" [China's Arctic Discourse Rights and Its Promotion Approach], in 北极地区发展报告2017 [*Report on Arctic Region Development 2017*], ed. Liu Huirong (Beijing: 社会科学文献出版社 [Social Sciences Academic Press], 2017), 80–101.

89 Yang Jian, 科学家与全球治理：基于北极事务案例的分析 [*Science Community and Global Governance: A Case Analysis on the Arctic Affairs*] (Beijing: 时事出版社 [Current Affairs Press], 2018): 251–252.

90 Yang Jian and Yu Hongyuan, "中国科学家群体与北极治理议程的设定：基于问卷的分析" [The Community of Chinese Scientists and the Agenda Setting of Arctic Governance – Based on an Analysis of the Questionnaires], 国际关系研究 [*Journal of International Relations*], no. 6 (2014): 37–49.

91 Rasmus Gjedssø Bertelsen, Li Xing and Mette Højris Gregersen, "Chinese Arctic Science Diplomacy: An Instrument for Achieving the Chinese Dream?" in *Global Challenges in the Arctic Region: Sovereignty, Environment and Geopolitical Balance*, eds. Elena Conde and Sara Iglesias Sánchez (Oxford: Taylor & Francis, 2016), 448.

92 Ibid., 448.

93 Ibid.

94 Sun and Wu, "北极治理新态势与中国的深度参与战略" [New Developments in Arctic Governance and China's Deepening Involvement Strategy], 78.

95 Liu Yang, "专家给中国北极战略提建议：可利用我遥感技术优势" [Experts to China's Arctic Strategy: Use Remote Sensing Technology Advantages], 环球时报 [*Global Times*], September 25, 2017, https://m.huanqiu.com/article/9CaKrnK5kOP.

96 Torbjørn Pedersen, "The Politics of Research Presence in Svalbard," *The Polar Journal* 11, no. 2 (2021): 413–426.

97 Yang, Jian, "以"人类命运共同体"思想引领新疆域的国际治理" [The Vision of a Community with a Shared Future for Mankind Will Guide International Governance in the New Frontiers], 人民网 [*People's Daily Online*], June 23, 2017, http://cpc.people.com.cn/n1/2017/0623/c191095-29358375.html.

98 Marc Lanteigne, "Have You Entered the Storehouse of the Snow? China as a Norm Entrepreneur in the Arctic," *Polar Record* 53, no. 269 (2017): 119.

99 Ibid., 118.

100 Gao Feng – Special Representative for Climate Change Negotiations of the Foreign Ministry of China, *Speech at the 2016 Arctic Circle Assembly*, Iceland (video), https://vimeo.com/189527301.

101 Zhang, "中国北极话语权及其提升路径研究" [China's Arctic Discourse Rights and Its Promotion Approach], 99.

102 Hong Nong, "Going Beyond the 'Original Inter-Arctic States': China Acting in the Arctic and Observing the Arctic Council," in *'Observing' the Arctic: Asia in the Arctic Council and Beyond*, eds. Chih Yuan Woon and Klaus Dodds (Cheltenham, UK and Northampton: Edward Elgar Publishing, 2020), 155.

103 N.L. Dobretsov and N.P. Pokhilenko, "Mineral Resources and Development in the Russian Arctic," *Russian Geology and Geophysics* 51 (2010): 98–111.

104 President of the Russian Federation, *Foundations of the Russian Federation State Policy in the Arctic for the Period up to 2035* (Moscow, Russia, 2020). Translated from Russian by Anna Davis and Ryan Vest, *Russia Maritime Studies Institute* (Newport, Rhode Island, 2020), https://digital-commons.usnwc.edu/rmsi_research/5/.

105 "Meeting on Arctic Zone Development," *President of Russia*, April 13, 2022, www.en.kremlin.ru/events/president/transcripts/68188.

106 For a brief overview of Sino-Russian relations since the end of the Cold War see Alexei D. Voskressenski, "China's Relations with Russia," in *China and the World*, ed. David Shambaugh (New York: Oxford University Press, 2020), 233–250.

107 "27th Regular Meeting of Russian and Chinese Heads of Government," *The Russian Government*, December 5, 2022, http://government.ru/en/news/47237/.

108 Evan S. Medeiros, "China's Strategic Straddle: Analysing Beijing's Diplomatic Response to the Russian Invasion of Ukraine," *China Leadership Monitor* 72 (2022): 8–12.

109 "Xi Wins Putin's Backing on Taiwan, Plays Peacemaker on Ukraine," *Nikkei Asia*, March 23, 2023, https://asia.nikkei.com/Politics/International-relations/Xi-wins-Putin-s-backing-on-Taiwan-plays-peacemaker-on-Ukraine2.

110 "Russia Expects Trade with China to Reach $200 Billion by 2024," *Reuters*, April 30, 2022, www.reuters.com/business/russia-expects-trade-with-china-reach-200-bln-by-2024-ifax-2022-04-30/.

111 Fu Ying, "How China Sees Russia," *People's Daily*, December 18, 2015, http://en.people.cn/n/2015/1218/c90000-8992446.html.

112 "Russian, Chinese Bombers Fly Joint Patrols Over Pacific," *AP News*, November 30, 2022, https://apnews.com/article/russia-ukraine-china-beijing-moscow-europe-3d694c61d318e083076b681cd913102d.

113 He Qisong and Ye Nishan, "Analysis of Space Cooperation Between China and Russia [中国与俄罗斯太空合作分析]," *Interpret: China* (2023, original work published August 2, 2021), https://interpret.csis.org/translations/analysis-of-space-cooperation-between-china-and-russia/.

114 "中俄举行第七轮北极事务对话" [China and Russia Held the Seventh Round of the Arctic Affairs Dialogue], 中华人民共和国外交部 [*Ministry of Foreign Affairs of the PRC*], April 14, 2021, www.mfa.gov.cn/web/wjb_673085/zzjg_673183/tyfls_674667/xwlb_674669/202104/t20210414_9176285.shtml.

115 "中华人民共和国和俄罗斯联邦关于发展新时代全面战略协作伙伴关系的联合声明（全文）" [Full Text of the Sino-Russian Joint Statement on Developing Comprehensive Strategic Partnership of Coordination in the New Era], *Xinhua*, June 6, 2019, www.xinhuanet.com/world/2019-06/06/c_1124588552.htm.

116 Malte Humpert, "Putin and Xi Discuss Further Deepening of Arctic Partnership," *High North News*, March 23, 2023, www.highnorthnews.com/en/putin-and-xi-discuss-further-deepening-arctic-partnership.

117 "中海油服圆满完成中国首次北极海域地震勘探作业" [COSL Successfully Completed China's First Seismic Exploration in Arctic Waters], 中海油田服务股份有限公司 [*COSL*], August 9, 2016, https://web.archive.org/web/20161007071459/www.cosl.com.cn/art/2016/8/9/art_14991_2401131.html.

118 Atle Staalesen, "Chinese Oilmen Make Big Discovery in Russian Arctic Waters," *The Barents Observer*, April 5, 2018, https://thebarentsobserver.com/en/industry-and-energy/2018/04/chinese-oilmen-make-big-discovery-russian-arctic-waters.

119 Malte Humpert, "Russian Mining Company Partners with China to Develop Massive Titanium Deposit in Arctic," *High North News*, February 6, 2023, www.highnorthnews.com/en/russian-mining-company-partners-china-develop-massive-titanium-deposit-arctic.

120 Malte Humpert, "Lacking Own Satellite Coverage Russia Is Looking to China for Northern Sea Route Data," *High North News*, March 30, 2023, www.highnorthnews.com/en/lacking-own-satellite-coverage-russia-looking-china-northern-sea-route-data.

121 "Potential for China-Russia Cooperation in the Russian Arctic Highlighted at Forum," *Global Times*, September 6, 2022, www.globaltimes.cn/page/202209/1274785.shtml.

122 "关于我们" [About Us], 中国人民大学重阳金融研究院 [*Chongyang Institute for Financial Studies at Renmin University of China*], n.d., http://rdcy.ruc.edu.cn/zw/gywm/RDCYjj/index.htm.

123 Chen Xiaochen and Zhang Tingting, 去欧洲，向北走：中俄共建"冰上丝绸之路"支点港口研究，研究报告第36期 [*Going to Europe, Head North: Research on the Establishment of Joint Sino-Russia Polar Silk Road Pivot Ports*, Research Report No. 36] (Beijing: 中国人民大学重阳金融研究院 [Chongyang Institute for Financial Studies at Renmin University of China], 2018).

124 Henry Tillman, Yang Jian and Egill Thor Nielsson, "The Polar Silk Road: China's New Frontier of International Cooperation," *China Quarterly of International Strategic Studies* 4, no. 3 (2018): 355.

125 Sergey Sukhankin, "Russia's Belkomur Arctic Railway Project: Hope, Illusion or Necessity?" *Eurasia Daily Monitor* 16, no. 102 (2019), https://jamestown.org/program/russias-belkomur-arctic-railway-project-hope-illusion-or-necessity/.

126 Humpert, "Russian Mining Company Partners with China to Develop Massive Titanium Deposit in Arctic."

127 Gao Qiao, "北极保护开发利用议题引各方热议"冰封之地"渐成关注热点" [The Issue of Protection, Development and Utilization of the Arctic Attracts Heated Discussions – The "Frozen Land" Is Gradually Becoming a Hot Spot], 人民日报 [*People's Daily*], October 23, 2021, http://paper.people.com.cn/rmrbhwb/html/2021-10/23/content_25884907.htm.

128 Hu Chenglin, "中俄首次北极联合科考 弥补我国历次北极考察区域不足" [China and Russia Conducted for the First Time a Joint Arctic Research Expedition Making Up for the Previous Deficiencies in Arctic Scientific Expeditions], 青岛大众网 [*Qingdao Public Network*], December 1, 2016, http://qingdao.dzwww.com/xinwen/qingdaonews/201612/t20161201_15224135.htm.

129 Zhao Ning, "中俄首次北极联合科考圆满结束 获丰硕科考成果" [The First Sino-Russian Arctic Joint Scientific Expedition Has Successfully Concluded and Achieved Fruitful Scientific Research Results], 中国海洋报 [*China Ocean News*], October 12, 2016, www.chinanews.com.cn/gn/2016/10-12/8028900.shtml.

130 Su Wanming, "2018年中俄北极联合科考取得多项成果" [The 2018 Joint Sino-Russian Arctic Scientific Expedition Achieved Many Results], *Xinhua*, October 30, 2018, www.gov.cn/xinwen/2018-10/30/content_5335979.htm.

131 Xie Chuanjiao, "Sino-Russian Expedition Provides Arctic Data," *China Daily*, October 31, 2018, www.chinadaily.com.cn/a/201810/31/WS5bd9016fa310eff30328591e.html.

132 Gao Yue, "联合科考领航"冰上丝绸之路" – 自然资源部第一海洋研究所对俄合作纪实" [Joint Scientific Research Navigates the Polar Silk Road – A Record of Cooperation Between the First Institute of Oceanography of the Ministry of Natural Resources and Russia], 中国自然资源报 [*China Natural Resources News*], March 23, 2023, www.mnr.gov.cn/dt/ywbb/202303/t20230323_2779077.html.

133 Pavel Devyatkin, "Russian and Chinese Scientists to Establish Arctic Research Centre," *High North News*, April 15, 2019, www.highnorthnews.com/en/russian-and-chinese-scientists-establish-arctic-research-center.

134 "哈尔滨工程大学获批首批"一带一路"联合实验室" [Harbin Engineering University Was Approved as the First Batch of the Belt and Road Joint Laboratories], 央广网 [*China National Radio Online*], June 21, 2019, https://brgg.fudan.edu.cn/articleinfo_1162.html.

135 "Far Eastern Federal University, Harbin Polytechnic University Establish a Russian-Chinese Polar Engineering Centre," *The Arctic*, September 29, 2016, http://arctic.ru/international/20160929/446559.html.

136 "第十届中俄北极论坛召开" [The 10th China-Russia Arctic Forum Was Held], 中国海洋大学海洋发展研究院 [*Institute of Marine Development of Ocean University*], September 9, 2021, http://hyfzyjy.ouc.edu.cn/2021/0909/c11432a346922/page.htm.

137 "中俄北极可持续发展合作与能源安全国际论坛在吉林大学召开" [The Sino-Russia International Forum on Arctic Sustainable Development Cooperation and Energy Security Was Held at Jilin University], 吉林大学东北亚研究中心 [*Northeast Asian Research Centre of Jilin University*], May 27, 2021, https://narc.jlu.edu.cn/info/1013/1681.htm.

138 Adam Perry MacDonald, "China-Russian Cooperation in the Arctic: A Cause for Concern for the Western Arctic States?" *Canadian Foreign Policy Journal* 27, no. 2 (2021): 197–198.

139 Deng, 北极安全研究 [*Arctic Security Studies*], 233.

140 Christopher Weidacher Hsiung, "China's Technology Cooperation with Russia: Geopolitics, Economics, and Regime Security," *The Chinese Journal of International Politics* (2021): 457.

141 Olga Alexeeva and Frederic Lasserre, "An Analysis on Sino-Russian Cooperation in the Arctic in the BRI Era," *Advances in Polar Sciences* 29, no. 4 (2018): 269–282.

142 Andrew Foxall, "The Sino-Russian Partnership in the Arctic," in *On Thin Ice? Perspectives on Arctic Security*, eds. Duncan Depledge and P. Whitney Lackenbauer (Peterborough, Canada: NAADSN, 2021), 82–90.

143 "Joint Statement of the Russian Federation and the People's Republic of China on the International Relations Entering a New Era and the Global Sustainable Development," *President of Russia*, February 4, 2022, http://en.kremlin.ru/supplement/5770.

144 Trym Eiterjord, "Amid Ukraine War, Russia's Northern Sea Route Turns East," *The Diplomat*, December 13, 2022, https://thediplomat.com/2022/12/amid-ukraine-war-russias-northern-sea-route-turns-east/.

145 Ibid.

146 Atle Staalesen, "Arctic Shipping and Energy on Putin's Agenda with Xi Jinping," *The Barents Observer*, March 21, 2023, https://thebarentsobserver.com/en/industry-and-energy/2023/03/arctic-energy-agenda-xi-meets-russian-pm-mishustin.

147 Solveig Glomsrød et al., "Arctic Economies Within the Arctic Nations," in *The Economy of the North 2015*, eds. Solveig Glomsrød, Gérard Duhaime and Iulie Aslaksen (Oslo: Statistics Norway, 2017), 44–47.

148 Heather A. Conley et al., *A New Security Architecture for the Arctic: An American Perspective* (Washington, DC: Centre for Strategic and International Studies, 2012), 18.

149 James Kraska, "Asian States in US Arctic Policy: Perceptions and Prospects," *Asia Policy* 18 (2014): 20–21.

150 "学院动态" [College News], 同济大学政治与国际关系学院 [*Tongji University School of Political Science and International Relations*], June 17, 2017, https://spsir.tongji.edu.cn/db/a9/c18896a187305/page.htm.

151 Peng Yining, "China Charting a New Course," *China Daily*, April 20, 2016, www.chinadaily.com.cn/china/2016-04/20/content_24679000.htm.

152 Yu Qiongyuan, "中国首次成功试航北极西北航道" [China for the First Time Successfully Navigated the Northwest Passage], *Xinhua*, September 7, 2017, http://news.xinhuanet.com/politics/2017-09/07/c_1121625642.htm.

153 "Alaska Sees Natural Gas New Catalyst for Closer Ties with China," *Xinhua*, September 25, 2017, www.chinadaily.com.cn/business/2017-09/25/content_32451940.htm.

154 The White House, *National Security Strategy* (Washington, DC, 2022), 23, www.whitehouse.gov/wp-content/uploads/2022/10/Biden-Harris-Administrations-National-Security-Strategy-10.2022.pdf.

155 Cited in Charles Hutzler, "House Committee Lays Out Case for China Threat," *The Wall Street Journal*, February 28, 2023, www.wsj.com/articles/house-committee-lays-out-case-for-china-threat-ad62c611.

156 "US Hegemony and Its Perils," *Ministry of Foreign Affairs of the PRC*, February 20, 2023, www.fmprc.gov.cn/mfa_eng/wjbxw/202302/t20230220_11027664.html.

157 Ryan Hass, "Beijing's Response to the Biden Administration's China Policy," *China Leadership Monitor* 71 (Spring 2022): 4, www.prcleader.org/hass-1.

158 See, for example, Thomas Nilsen, "US Navy Will Build Airport Infrastructure in Northern Norway to Meet Upped Russian Submarine Presence," *The Barents Observer*, April 16, 2021, https://thebarentsobserver.com/en/security/2021/04/us-navy-build-airport-infrastructure-northern-norway-meet-increased-russian.

159 Megan Eckstein, "US, UK Surface Warships Patrol Barents Sea for First Time Since the 1980s," *USNI News*, May 4, 2020, https://news.usni.org/2020/05/04/u-s-u-k-surface-warships-patrol-barents-sea-for-first-time-since-the-1980s.

160 Malte Humpert, "US Coast Guard Awards Contract for New Polar Class Icebreaker," *High North News*, February 9, 2023, www.highnorthnews.com/en/us-coast-guard-awards-contract-new-polar-class-icebreaker.

161 The White House, *National Strategy for the Arctic Region* (Washington, DC, 2022), 6, www.whitehouse.gov/wp-content/uploads/2022/10/National-Strategy-for-the-Arctic-Region.pdf.

162 Yang, *Arctic Governance and China's Engagement*, 144.

163 Li Yun, "相关国家强化北极地区军事部署，冷战思维死灰复燃" [Countries Have Intensified Military Deployment in the Arctic, Reviving the Cold War Mentality], 解放军报 [*PLA Daily*], June 9, 2022, www.81.cn/wj_208604/10161736.html.

164 Lin Yuan, "美新版北极战略加剧极地博弈" [The New US Arctic Strategy Intensifies the Polar Games], 解放军报 [*PLA Daily*], October 14, 2022, www.81.cn/gfbmap/content/2022-10/14/content_325714.html.

165 Xiao Yang, "竞争性抵制：美国对冰上丝绸之路的拒阻思维与战略构建" [Competitive Resistance: American Resistance to the Polar Silk Road and Strategic Construction], 太平洋学报 [*Pacific Journal*] 27, no. 7 (2019): 66–75.

166 Liu Dan, "中国北极威胁论：现状、原因与影响" [The China Arctic Threat Theory: Present Situation, Causes and Impacts], 西部学刊 [*The Western Journal*] 2 (2022), www.fx361.com/page/2022/0425/10274442.shtml.

167 For example, see: Lu Zhongwei, "谁在给北极披战袍" [Who Is Wrapping the Arctic in Armour], 文汇报 [*Wen Hui Bao*], May 11, 2019, www.sohu.com/a/313490445_618422.

168 Tom Daly and Jeff Lewis, "Canada Rejects Bid by China's Shandong for Arctic Gold Mine on Security Grounds," *Reuters*, December 22, 2020, www.reuters.com/article/us-tmac-resources-shandong-gold-idUSKBN28W18R.

169 "Foreign Ministry Spokesperson Hua Chunying's Regular Press Conference on April 20, 2016," *The Ministry of Foreign Affairs of the PRC*, April 20, 2016, https://web.archive.org/web/20160426120112/www.fmprc.gov.cn/mfa_eng/xwfw_665399/s2510_665401/t1357177.shtml.

170 Ken S. Coates and Carin Holroyd, "Europe's North: The Arctic Policies of Sweden, Norway, and Finland," in *The Palgrave Handbook of Arctic Policy and Politics*, eds. Ken S. Coates and Carin Holroyd (Cham, Switzerland: Palgrave Macmillan, 2020), 283–284.

171 Ibid., 283.

172 "徐绍史、刘赐贵拜会格陵兰自治政府总理库皮克" [Xu Shaoshi and Liu Cigui Visit Greenland's Prime Minister Kuupik Kleist], 国土资源部 [*Ministry of Land and Resources of the PRC*], April 27, 2012, www.gov.cn/gzdt/2012-04/27/content_2124897.htm.

173 "Successful Visit by Chinese Head of State in Iceland," *Iceland Review*, April 23, 2012, www.icelandreview.com/news/successful-visit-chinese-head-state-iceland/.

174 "Free Trade Agreement Between Iceland and China," *Government Offices of Iceland*, n.d., www.government.is/topics/foreign-affairs/external-trade/free-trade-agreements/free-trade-agreement-between-iceland-and-china/.

175 "China, Finland Confirm Establishment of Future-Oriented New-Type Cooperative Partnership," *Xinhua*, April 5, 2017, http://news.xinhuanet.com/english/2017-04/05/c_136185070.htm.

176 "Joint Declaration Between the Republic of Finland and the People's Republic of China on Establishing and Promoting the Future-Oriented New-Type Cooperative Partnership," *The President of the Republic of Finland*, April 5, 2017, www.presidentti.fi/en/news/joint-declaration-between-the-republic-of-finland-and-the-peoples-republic-of-china-on-establishing-and-promoting-the-future-oriented-new-type-cooperative-partnership/.

177 Thomas Nilsen, "China Wanted to Buy Airport in Lapland for North Pole Climate Research Flights," *The Barents Observer*, March 4, 2021, https://thebarentsobserver.com/en/climate-crisis/2021/03/china-wanted-buy-airport-lapland-north-pole-climate-research-flights.

178 Deng, 北极安全研究 [*Arctic Security Studies*], 234.

179 Camilla T.N. Sørensen, "Chinese Investments in Greenland: Promises and Risks as Seen From Nuuk, Copenhagen and Beijing," in *Greenland and the International Politics of a Changing Arctic: Postcolonial Paradiplomacy Between High and Low Politics*, eds. Kristian Søby Kristensen and Jon Rahbek-Clemmensen (New York: Routledge, 2017), 87.

180 Atle Staalesen, "Barents Town Envisions Arctic Hub with Link to China," *The Barents Observer*, February 6, 2018, https://thebarentsobserver.com/en/arctic/2018/02/barents-town-envisions-arctic-hub-link-china.

181 Erik Matzen, "Denmark Spurned Chinese Offer for Greenland Base Over Security: Sources," *Reuters*, April 6, 2017, www.reuters.com/article/us-denmark-china-greenland-base-idUSKBN1782EE.

182 Drew Hinshaw and Jeremy Page, "How the Pentagon Countered China's Designs on Greenland," *The Wall Street Journal*, February 10, 2019, www.wsj.com/articles/how-the-pentagon-countered-chinas-designs-on-greenland-11549812296.

183 Xiao Yang, "冰上丝绸之路的战略支点 – 格陵兰"独立化"及其地缘价值" [The Strategic Fulcrum of the Silk Road on Ice – Greenlandic Independence and Its Geographical Value], 和平与发展 [*Peace and Development*] 6 (2017): 108–123.

184 The Police Security Service, *National Threat Assessment 2023* (Oslo, 2023), 12, https://web.archive.org/web/20230224111537mp_/www.pst.no/globalassets/ntv/2023/ntv_2023_eng_web.pdf.

185 Ibid., 12.

186 Matti Puranen and Sanna Kopra, "Finland and the Demise of China's Polar Silk Road," *China Brief* 22, no. 24 (2022), https://jamestown.org/program/finland-and-the-demise-of-chinas-polar-silk-road/.

6   Conclusion and Implications

In the 21st century, the PRC has emerged as a significant global power with the capacity, resources, and ambition to reshape the international order in line with its own interests, and preferences. China's core interests are primarily located in East Asia, particularly those connected to what the CCP views as part of its territorial integrity, that is, Taiwan and the South China Sea. However, as Yoshihara and Holmes remind us, China's dreams and ambitions are not bounded by this geographical limitation.[1] The notion of a greater Chinese involvement in global affairs became significantly more prominent under the leadership of Xi Jinping, who began implementing his vision of the Great Rejuvenation of the Chinese Nation, ushering China into a New Era of development. In accordance with this grand plan, the CCP is restructuring the Chinese economy towards high-quality development with a focus on innovation, advanced manufacturing, and self-reliance, all underpinned by a strong emphasis on science and technology advancements. This is accompanied by sustained efforts to modernise the Chinese armed forces, aiming to transform them into a world-class military. To achieve this, the CCP has spearheaded the strategy of military-civil fusion which seeks to integrate and leverage the synergies between the military and civilian sectors, such as in hardware development. To address various domestic and international political, social, economic, military, and ideological risks that could hinder the PRC's modernisation, the CCP adopted a Comprehensive National Security Outlook in 2014, encompassing 16 types of security and emphasising the paramount importance of security-related matters within the Party-state and society. On the global stage, the PRC is increasingly vocal about perceived injustices in the American-led world order and adopts a more assertive stance in defending what it deems as its core interests. This assertiveness is also reflected in the various ideas and initiatives, such as the BRI and the Community of Shared Future for Mankind, which the PRC is proposing as solutions to the issues faced by the international community in the hopes of increasing its discourse power while simultaneously challenging the dominant role of the US in global affairs.

A prominent aspect of Xi Jinping's leadership in China is the country's relentless pursuit of deeper involvement in what is referred to as the strategic new frontiers, namely, the polar regions, deep sea, cyberspace, and outer space. These frontiers are seen in China as arenas of great power competition for resources and

DOI: 10.4324/9781003295112-6

international influence, which can contribute to a state's economic growth but currently lack effective governance structures. The CCP's stated goals of transforming the PRC into a polar great power or a space great power demonstrate the country's readiness to exert a greater role in and influence over these new frontiers. To facilitate this, the PRC possesses an extensive state bureaucracy comprising government ministries and departments, SOEs, research institutes, and universities.

This book focused on a specific area within the strategic new frontiers: the Arctic region. This region has garnered global attention due to environmental changes, its abundant resources, and increasing remilitarisation, including from states well outside the region such as China. It attempted to locate the region within the PRC's national strategy of the Great Rejuvenation of the Chinese Nation, as seen from the Chinese perspective and portrayed in official PRC policy documents, and Chinese elite discourses. In this chapter, the first section summarises the preceding chapters, which focused on Chinese Arctic science, security, and governance, highlighting connections to China's national strategy and its impacts on the country's comprehensive national power. The second section examines the implications of China's deepening interests in the Arctic for its relations with the US and the broader structures guiding regional governance.

The Arctic and China's National Rejuvenation

Under General Secretary Xi's leadership in the New Era, the PRC has intensified its engagement with the Arctic region. The country published an Arctic White Paper in 2018 which identified China as a near-Arctic state and emphasised its role as an important Arctic stakeholder. The white paper acknowledged that China's climate system and, by extension, its economic development, were influenced by Arctic changes. China has also expanded its scientific involvement in the region, which now includes two regional research stations, two research icebreakers, and a number of domestic research institutes and universities dedicating resources to studying the Arctic. Chinese SOEs have invested in resource and shipping development projects above the Arctic Circle and continue to explore opportunities for greater presence in the region. The Chinese Party-state has also increased its participation in regional governance forums and conferences, such as the AC and the Arctic Circle Assembly, using these platforms to share its perspectives on regional affairs. Moreover, the PRC has deepened its ties with one of the key Arctic states, Russia. It is important to recognise that China is no longer a newcomer to Arctic regional affairs, but a great power seeking to exert influence in the Arctic region.[2]

Since the late 1990s, China's Arctic engagement has prioritised scientific research, knowledge development, and understanding of the region's changes and their impacts beyond the Arctic, particularly on the Chinese mainland. Chinese Party-state institutions, including the PRIC, have established research stations in Svalbard and Iceland, conducted marine research expeditions using Chinese research icebreakers Xuelong and Xuelong 2, and deployed a multitude of instruments, such as unmanned underwater vehicles, buoys, meteorological stations, and Earth-orbiting satellites, providing crucial information and data to Chinese scientists. These efforts, in collaboration with other Chinese research and academic institutions, aim to advance China's plans to establish a multidimensional Arctic research and monitoring system.

As outlined in Chapter 3, Chinese engagement in Arctic science activities has connected China to international cooperative networks and legitimised its presence and interest in the region. It has also helped the PRC to assert its near-Arctic identity and obtain observer status in the AC. At the national level, within the context of China's strategy of national rejuvenation, and the security domain in particular, Chinese Arctic science and knowledge development help the CCP to safeguard its ecological security by providing accurate data, observations, and predictions on the impacts of Arctic changes on the Chinese mainland. This enables the CCP to develop contingencies and avoid disruptions, including potential economic instability caused by extreme weather events resulting from Arctic environmental changes.[3]

As China's national rejuvenation is closely tied to the modernisation of the PLA and its transformation into a world-class fighting force, the Arctic region assumes growing importance in terms of hard security. It seems that at present, the Chinese armed forces are developing polar capabilities and enhancing their Arctic situational awareness. This effort, as explained in Chapter 4, is not aimed at military domination of the Arctic but rather to have the capabilities to operate effectively in the harsh Arctic environment if the CCP wants to field a true blue-water navy and become a peer competitor to the US military, claiming the status of a maritime great power. Through its military-civil fusion strategy, the Chinese Party-state can leverage the knowledge gained from extensive scientific engagement in the Arctic to advance these goals. Moreover, building Arctic-related military capabilities enhances China's military security by developing competence that will allow the Chinese armed forces to defend its overseas interests and, to some extent, secure its northern flank in cooperation with Russia, thereby safeguarding China's sovereignty.

China also recognises the value of the Arctic region in terms of natural resources, particularly energy and green technologies, as well as the potential for shorter and faster shipping routes. In this regard, Chinese SOEs have invested in resource development projects in the Russian Arctic and shown considerable interest in the mineral potential of Greenland. In terms of shipping, China's COSCO is one of the largest non-Russian operators on the NSR, actively exploring opportunities for transit and destination shipping. As outlined in Chapter 4, these endeavours contribute to safeguarding China's economic security, particularly in diversifying energy imports and the development of alternative shipping routes that bypass the Indian Ocean to reach the European market. These efforts are crucial for sustaining the continued development of the Chinese state and society.

China's engagement in the Arctic across scientific, military, and commercial domains has the potential to drive technological innovation within the country. Each of these areas requires specialised knowledge, unique technologies, and a deep understanding of the Arctic environment. China's involvement in the Arctic allows it to test and develop materials, software, manufacturing techniques, and ultimately, the required specialised technologies. The construction of China's second research icebreaker, the Xuelong 2, serves as a notable example of this process. Developing such capabilities is no easy task, as seen in the challenges faced by the US in constructing its own fleet of polar class icebreakers due to a lack of recent experience in this field – the last US heavy icebreaker was constructed in the 1970s. The pursuit of high-tech innovations, such as unmanned systems and equipment for resource extraction in extreme climate conditions align with Xi Jinping's vision of restructuring the Chinese economy towards high-quality development. It also reflects the CCP's goals of reducing dependence on foreign technology and software, particularly from Western sources, to ensure technological security. With increased geopolitical competition between the US and China and calls for decoupling from some Western political factions, achieving technological self-reliance and fostering in-house innovation have gained greater urgency in China. Additionally, China's Arctic scientific, commercial, and even military engagement can also contribute to the development of the country's human capital. Investment projects involving Chinese participation above the Arctic Circle foster the growth of specific managerial and logistical knowledge and skills. Polar-related research fields attract talent which the PRC can use to further advance its understanding of the region. These experts can then represent China in Arctic-specific international collaborative projects and governance forums. Moreover, its engagement can help cultivate a competent workforce well-versed in advanced manufacturing techniques.[4] These efforts align with Xi's strategy of nurturing a high-quality workforce and positioning the PRC as a global hub of professional talent.

China's involvement in the Arctic is also linked to its political security objectives, aimed at maintaining regime and social stability. Chinese provinces, particularly those experiencing slower economic development than the national average (such as the ones located in the Chinese northeast) seek to capitalise on the opportunities presented by the Arctic region through China's foreign policy initiatives, including the BRI. If successful, these provinces could develop, potentially in cooperation with Russia, new transportation corridors that would open up local economies to maritime trade, some of which could lead through the Arctic region, spurring economic growth, job creation, and retaining young talent within these provinces. Furthermore, the CCP could leverage its achievements in Arctic scientific research, infrastructure, investments, and deployments to enhance its domestic legitimacy and satisfy nationalist demands among Chinese for international prestige and respect.

In terms of Arctic regional governance, as outlined in Chapter 5, the PRC has clearly expressed its intention not to be absent from the region, perceiving it as a global space open to both regional and extra-regional states. While officially upholding the Arctic governance system, the PRC subtly suggests that there is room for improvement. This viewpoint is much more pronounced within Chinese academic circles, which openly criticise Arctic governance. In particular, they highlight the dominant position of Arctic states in local governance structures and their perceived attempts at excluding extra-regional actors from participating in regional affairs. Chinese analysts argue for a more inclusive regional governance system that considers the interests and demands of states beyond the eight Arctic governments. In line with this narrative, some in the Chinese academic community propose applying Xi's vision of a Community of Shared Future for Mankind to the Arctic, using the PSR as a means to achieve this. As China's power and influence continue to grow both globally and regionally, the PRC will have the strength to enhance its discourse power and, as a norm entrepreneur, advocate for a more inclusive approach to regional affairs.

Simultaneously, there are various actors, organisations, and institutions within the Chinese Party-state system, at both the national and subnational levels, involved in China's multifaceted engagement in the Arctic. The CCP can leverage these actors to attain its domestic and international objectives related to the rejuvenation of the Chinese nation. The Party Centre, represented by the top leadership in Beijing, including the Politburo and its Standing Committee, is being strategically aware of the region and socialised to its policy importance. Xi's goal of transforming China into a polar great power drives this socialisation process, supported by high-level visits to Arctic states and policy dialogues with their leaders. Sino-Russian interactions, in particular, now incorporate an Arctic dimension, with references to Arctic development featuring in high-level declarations issued after joint meetings and bilateral visits. At the State Council level, several ministries and commissions, such as the Ministry of Natural Resources and the Ministry of Foreign Affairs, actively contribute to the development and implementation of Arctic policy within the broad parameters set by the Party leadership. These government organisations serve as official representatives of the Chinese Party-state in various Arctic multilateral and bilateral settings, including the Arctic Council and the Arctic Circle Assembly.

Beyond the central and ministerial-level organisations, actors linked to the Party-state, such as commercial and state-owned enterprises, research institutes, and universities, physically represent China's presence in the Arctic region. These entities establish on-site infrastructure, conduct ground-level research, engage in science diplomacy, and participate in international cooperative projects. Notably, the PRIC and its activities in Iceland exemplify this engagement, including the establishment of CNARC and the joint construction of the CIAO to study Arctic atmospheric phenomena. These actors also invest in Arctic hydrocarbon and mineral extraction projects, utilise Arctic shipping routes (e.g., CNPC's participation in Arctic LNG projects or COSCO's use of the NSR) and develop equipment for logistically and technologically challenging operations above the Arctic Circle. Through their activities in the Arctic region, they acquire specialised knowledge, expertise, and managerial skills. Additionally, these subnational entities can form strategic partnerships, often in collaboration with Chinese associations with relevant expertise, to enhance their efficiency and success in the Arctic region. As entities linked to the Chinese Party-state, they serve as knowledge producers, collectors, and communicators, contributing policy information to China's decision-makers, sharing Chinese ideas and concepts internationally, and offering potential solutions to perceived deficiencies in Arctic governance. It is through these domestic actors that the Party-state has established its presence in the region, built institutional capacity, and intensified engagement, especially in the past decade, ultimately leading to proclamations of China as a near-Arctic state and its aspirations of becoming a polar great power.

Consequently, the PRC's active engagement with the Arctic, as seen through the prism of Xi's Comprehensive National Security Outlook and the strategy of national rejuvenation, enhances the synergy between development and security in the PRC. This sustained focus on ensuring security and development – especially in areas that the Party considers critical to its future development and in which it seeks to attain self-sufficiency, such as technology and advanced manufacturing – aims to strengthen Chinese national power and state capacities. These advancements will allow the CCP to harden the Party-state system against external shocks – such as supply chain issues or external military threats – ensure the survivability of the CCP regime and create favourable conditions for further expanding Chinese influence into regions beyond China's immediate neighbourhood. Expansion can certainly imply imperialism; however, borrowing from Fareed Zakaria, it can also be seen as an activist or assertive foreign policy, characterised by heightened attention to international events, involvement in multilateral arrangements, and great power diplomacy.[5] In other words, Xi's national strategy of the great rejuvenation is creating conditions in China to develop necessary economic, scientific, technological, and human capacities, as well as the state's abilities to harness these resources in support of the Party's security and foreign policy initiatives.[6]

Looking ahead, under Xi's leadership, we can expect continued PRC engagement with the Arctic across political, scientific, commercial, and military domains. While some international analysts have noted signs of a Chinese "tactical retreat" in the Arctic, particularly in states allied with the US,[7] China's strategy of national rejuvenation demands an increased role in global affairs to expand its international influence. This also aligns with expectations of the Chinese population: a 2023 public opinion survey, conducted by the Centre for International Security and Strategy at Tsinghua University, China's top and internationally recognised university, found that 78% of respondents expect China's posture on foreign strategies to be more proactive.[8] Additionally, Chinese national images of the country being a near-Arctic state, an important Arctic stakeholder, and an Arctic affairs active participant, builder, and contributor, seem to be internalised in Chinese official and academic discourses. Such national self-images are likely to have a constitutive effect on China's conduct toward the region,[9] effectively becoming a part of its identity and a part of the messaging the CCP delivers to the international community. This in turn requires the PRC to remain engaged in Arctic affairs and further conditions it to contribute to regional governance, including the possibility of rule-making.

Implications for Inter-State Relations and Arctic Regional Structures

The PRC's continued interests and presence in the Arctic region, coupled with its broader international assertiveness, will further intensify the competitive nature of US-China relations, which is characterised by the dynamics of the security dilemma.[10] Originally introduced during the Cold War, the security dilemma describes a self-perpetuating cycle of power accumulation by states in a world defined by anarchy, uncertainty, and self-help. In such a world, when one state acquires power and capabilities for its own security and survival, other states perceive it as a threat to their own security. This leads them to acquire more power and capabilities to counter the power of the other state, generating a cycle of power competition, which can lead to a downward spiral of rivalry, arms races, and even military conflict.[11] While the security dilemma, as an analytical concept, has traditionally been applied to explain war, conflict, and issues pertaining to military affairs among states (and other groups), its dynamics can also be applied to fields such as technology and economics, which can be seen playing out in the Arctic between the US and China.

In light of the PRC's pursuit of military-civil fusion, Chapter 4 discussed the potential strategic use of Chinese Arctic science, particularly in developing the PLAN's Arctic situational awareness and dual-use technologies like UUVs. This may lead to the development of the so-called dual-use security dilemma. In a relationship characterised by extremely low levels of trust, such as that between the US and China, strategic science and dual-use technologies can add an additional layer of uncertainty to the already fraught relationship, as states are unsure about the intentions of their opponents in the development, acquisition, and production of these technologies, and whether or not these will be used for military or civilian purposes.[12] This uncertainty impacts the overall state of bilateral relations. An example of this dynamic on the global level can be seen in how the US and China accuse each other of being the bigger cybersecurity threat to each other and the world.[13] In the Arctic, the development of technologies for continuous observation of the region and its changing environment, including unmanned systems, ice stations, and satellites, can reflect this dynamic. While these technologies can serve purely scientific purposes, such as studying the Arctic seabed, they also have dual-use applications, such as submarine warfare.[14] Therefore, when China deploys these systems in the Arctic, this can generate the aforementioned dual-use security dilemma dynamics. The US military has already expressed concerns about the use of such science and systems in the Arctic by Chinese actors for military purposes, and the Chinese likely have similar concerns about American activities in waters near their homeland.

Similar dynamics can be observed in the economic realm. Recent moves by General Secretary Xi to strengthen the CCP's control over the Chinese economy in general and SOEs in particular (in response to perceived domestic and external threats) have raised concerns among developed economies, including the US, about the true intentions of these state-owned actors.[15] Pearson, Rithmire, and Tsai argue that the expansion of CCP cells in Chinese companies, the implementation of national security laws (some of which instruct Chinese organisations and individuals to assist in national intelligence work), and the military-civil fusion strategy integrating economic and defence development have further blurred the lines between the Party and the firm (as an autonomous commercial entity). These factors have generated insecurity in other states, leading them to adopt countermeasures to respond to these perceived economic uncertainties such as stricter investment screening of Chinese companies.[16] However, the PRC views these countermeasures as economic coercion, prompting the Chinese to respond accordingly. Port infrastructure involving Chinese capital and firms has also been a subject of concern; Kardon and Leutert argue that Chinese SOEs can potentially use their global port positions (whether through direct ownership, construction or other forms of investment) as an alternative source of intelligence collection and logistical support for military forces in peacetime "without the geopolitical consequences that dedicated overseas bases would trigger."[17] Given these dynamics, it is relatively easy to see why the US is increasingly anxious about the activities of Chinese companies in the US and globally. This unease was evident in the Arctic region when Chinese commercial entities attempted to invest in, acquire, or develop port infrastructure in Greenland. As indicated in Chapter 5, Danish authorities, seemingly influenced by US intervention, eventually prevented Chinese companies from participating in these projects. Similar concerns arose in Iceland and Scandinavia when Chinese actors expressed interest in purchasing land or investing in port development, such as the Lysekil port in Sweden, ultimately leading to the cancelation of these projects.[18]

In contrast, the PRC's relations with Russia appear to be moving towards greater cooperation and coordination, including in the Arctic region. Even before the Russian invasion of Ukraine in 2022, political leaderships in both countries seemed to be confident that the other side would not undermine the cooperative nature of their bilateral relationship by joining the West.[19] In 2020, Li Hui, former ambassador to Russia and current Chinese envoy for the settlement of the war in Ukraine, emphasised that Sino-Russian relations had entered a new historical phase. He described the two countries as fellow travellers in the shared endeavour of constructing a Community of Shared Future for Mankind.[20] Despite China's growing national power, which should trigger security dilemma dynamics in Russia, some Russian scholars argue that Russia has no reason to be concerned. Alexander Lukin, a Russian political scientist, and an international relations expert from HSE University in Moscow, maintains in his analysis of Sino-Russian relations that both countries view each other as valuable partners and share a common worldview that seeks a less Western-centric world order. Consequently, there is no need for mutual wariness.[21] The Russian war against Ukraine, it would seem, did not alter this assessment. On the contrary, it accelerated Russia's reorientation towards Asian nations, particularly China. While there are some academics in the PRC admitting that Russia's actions in Ukraine present challenges to Chinese foreign policy principles, there seems to be a general understanding that maintaining closer ties with Russia is currently preferable (but not forming an alliance).[22] Meanwhile, Russia is increasingly dependent on the Chinese market for its imports and exports as it faces isolation from the European market. Bilateral trade between China and Russia saw a significant rise in the first five months of 2023, with Sino-Russian trade increasing by 40.7% and Russian imports from China by 75.6%, leading one expert to remark that the "Sinification of Russia's economy is on steroids."[23] Similar trends can now be observed in the Arctic dimension of the Sino-Russian partnership. Russian political leaders have expressed their willingness to afford China a greater role in the development of the NSR and hydrocarbon projects in the Russian Arctic. Furthermore, Russia has shown interest in Chinese technology and solutions to bolster "the technological sovereignty" of both nations.[24] According to Gabuev, China is now in an advantageous position vis-à-vis Russia to advance its national interests, including acquiring Russian energy resources at discounted prices, expanding its market for Chinese technology and standards and even gaining preferential access to the Russian Arctic.[25]

The intensifying global antagonism between the United States and China, coupled with the prospect of ongoing amicability between China and Russia, will have implications for the future development of Arctic regional arrangements. Many international political elites, analysts, and academics point to the division of the global state system into two competing blocs: one centred around the US and its allies, and the other around China and its network of friendly nations. UN Secretary-General António Guterres expressed concerns about this division in 2019, warning of

> the world splitting in two, with the two largest economies on earth creating two separate and competing worlds, each with their own dominant currency, trade, and financial rules, their own internet and artificial intelligence capacities, and their own zero-sum geopolitical and military strategies.[26]

This perceived bifurcation of global politics has led many to draw parallels with previous US-USSR confrontations, characterising it as a new Cold War. Zhao Minghao of Fudan University points out that this new Cold War "is characterised by intense competition among the major powers for control of global commons, such as the Internet and outer space, wherein contests are largely associated with controlling connectivity rather than occupying territory."[27] Various predictions have been put forth regarding what kind of an international order will emerge given the shifting global power dynamics. For instance, Mearsheimer suggests the formation of two thick bounded orders – one American and one Chinese – marked by intense security competition between them.[28] On the other hand, Yan Xuetong envisions an international order of uneasy peace defined by independent digital systems, proxy wars, double-dealing, and technological decoupling.[29]

Considering the way global dynamics manifest themselves at the regional levels, it is likely that a similar division will occur in the Arctic. Building on Mearsheimer's conceptual framework, Rasmus Bertelsen argues that the US and China will eventually establish their bounded orders in the Arctic, similar to the way the US and USSR did in the past. The American bounded order will encompass the North American and Nordic states, while the Chinese, with Russia's acquiescence, will establish a joint Sino-Russian Arctic bounded order.[30] Such an arrangement is plausible, and if it materialises, it will be underpinned by Chinese technology and standards. As mentioned earlier, following the 2022 Russian invasion of Ukraine, substantial contacts between the so-called West and Russia, including in the Arctic, have diminished. The West actively seeks to isolate Russia economically, militarily, and technologically. In response, Russia sees closer cooperation with Asian states, particularly China, as a pragmatic choice to counterbalance these isolation attempts. However, this situation creates opportunities for China to expand its presence in the Russian market and supplement Western brands, products, and technological solutions with Chinese alternatives. If the current efforts to isolate Russia persist, coupled with China's previous experiences in resource extraction projects in Russia, China may feel confident to fully step in and provide all of the financial, technological, and logistic support the Russians need to continue developing their Arctic zone. This scenario would result in a situation in which half of the Arctic region would be nominally under Russian control but heavily dependent on Chinese investments and technology.

However, these divisions will have implications for certain arrangements that facilitate Arctic governance, such as those concerning Svalbard and the AC. Under the Svalbard Treaty, Norway has full and absolute sovereignty over the archipelago but must grant nationals of treaty member states equal rights to those of Norwegian nationals. The Norwegian government, in the spirit of the treaty, also facilitates international research activities in Svalbard by providing facilities and equipment in Ny-Ålesund, which it considers as one large Norwegian research base open to international participation.[31] However, not all parties, including the PRC, share the same view. Pedersen notes that Chinese representatives expressed dissatisfaction when presented with a draft of Norway's research strategy for Ny-Ålesund, disagreeing with its characterisation as a Norwegian research base. They apparently demanded more autonomy from the Norwegian government by "requesting an international decision-making process considering the special features of Ny-Ålesund and Spitsbergen as a whole" and stating that they "would not accept that our station would be referred to as a certain building of facility belonging to the so-called Ny-Ålesund Research Station."[32] The growing discord between China and the West could further exacerbate the divergence in interpreting established international laws and agreements. Moreover, Russia has recently declared its readiness to establish a new research station on Svalbard in cooperation with the rest of the BRICS countries (China, Brazil, India, and South Africa).[33] While it remains unclear whether China would participate in such a project, given its existing research station in Ny-Ålesund, and how Norway would respond, but such possibility should not be discounted considering the close relationship between China and Russia. There are also significant uncertainties surrounding the future of the AC. Although Norway assumed the chairmanship in May 2023 without substantial controversies, subsequent Russian statements questioned the fate of the AC if strained relations between Russia and the West persisted.[34] The Chinese, through their Special Representative for Arctic Affairs, have already emphasised the difficulty of supporting the Council's work in the absence of Russia.[35] These developments could hinder progress in regional cooperation needed to address issues of common (intra and extra-regional) concern, including environmental protection, economic development, and cross-border interactions, thereby fuelling calls for reforming regional governance arrangements.

China's involvement in the Arctic region, with support from Russia, can be seen as the final step in the broader process of greatly expanding the PRC's influence over Eurasia and its surrounding maritime areas. Geo-economically, the PRC has been spearheading the BRI whose three principal physical routes, the 21st Century Maritime Silk Road, the Economic Silk Road Belt, and the Polar Silk Road envelop Eurasia, opening neighbouring regions to Chinese goods, investments, and technology. The successful implementation of the BRI will strategically position the PRC at the centre of the continent and greatly improve its external security environment.[36] Politically, this will provide the PRC with the necessary material and discursive power to promote its solutions to the perceived shortcomings of a US-dominated world and push for its visions of a different regional and global order. Chinese political and academic elites, including the top leadership of the CCP, have emphasised the need to establish a Community of Shared Future for Mankind to tackle global issues. China's active presence in the Arctic, including participation in regional governance structures and conferences, offers an additional channel for Chinese actors to infuse regional narratives with Chinese ideas and proposals. It is worth noting that not everyone opposes China playing a larger role in Asian politics. Some Russian scholars are already exploring potential synergies between Chinese and Russian initiatives, such as the Eurasian Economic Union, and contemplating the idea of a non-Western Greater Eurasian community.[37] Beyond the geo-economic and political considerations, China's involvement in the Arctic provides valuable knowledge, expertise, and experience that will further enhance Chinese maritime and space capabilities. It is the ability of the US to study, develop, and field capabilities to control these commons (the sea and space) that enables its hegemony.[38] As such, through the Arctic, the PRC might gain a leading position on the Eurasian continent, returning China to what many in the PRC believe is the country's rightful place and fulfilling China's journey towards national rejuvenation.

Notes

1 Toshi Yoshihara and James R. Holmes, *Red Star Over the Pacific: China's Rise and the Challenge to US Maritime Strategy*, second edition (Annapolis: Naval Institute Press, 2018), 296.

2 Camilla T.N. Sørensen, "China Is in the Arctic to Stay as a Great Power: How China's Increasingly Confident, Proactive and Sophisticated Arctic Diplomacy Plays into Kingdom of Denmark Tensions," in *Arctic Yearbook 2018 – Arctic Development, In Theory & In Practice*, eds. Lassi Heininen and Heather Exner-Pirot (Akureyri, Iceland: Northern Research Forum, 2018), 1–15, https://arcticyearbook.com/images/yearbook/2018/China-and-the-Arctic/3_AY2018_Sorensen.pdf.

3 Linda Jakobson, "China's Security and the Arctic," in *Routledge Handbook of Chinese Security*, eds. Lowell Dittmer and Maochun Yu (London and New York: Routledge, 2015), 158.

4 Adopted from Dean Cheng, "For China, Space Is Both Substance and Symbol," *Space News*, January 25, 2021, https://spacenews.com/op-ed-for-china-space-is-both-substance-and-symbol/.

5 Fareed Zakaria, *From Wealth to Power: The Unusual Origins of America's World Role* (Princeton: Princeton University Press, 1999), 5.

6 Neoclassical realists in particular refer to the state's ability to mobilise resources for supporting policy initiatives as state power (or national political power) and view it as a

function of national power. See Norrin M. Ripsman, Jeffrey W. Taliaferro and Steven E. Lobell, *Neoclassical Realist Theory of International Politics* (New York: Oxford University Press, 2016), 145.

7 For "tactical retreat" reference see: Camilla T.N. Sørensen and Christopher Weidacher Hsiung, "The Role of Technology in China's Arctic Engagement: A Means as Well as an End in Itself," in *Arctic Yearbook 2021: Defining and Mapping the Arctic: Sovereignties, Policies and Perceptions*, eds. Lassi Heininen, Heather Exner-Pirot and Justin Barnes (Akureyri, Iceland: Arctic Portal, 2021), 11, https://arcticyearbook.com/images/yearbook/2021/Scholarly-Papers/11_AY2021_Sorensen_Hsiung.pdf.

8 Centre for International Security and Strategy, *2023 Public Opinion Polls: Chinese Outlook on International Security* (Beijing: Tsinghua University, 2023).

9 For the role of images in China's foreign policy see: Wang Hongying, "National Image Building and Chinese Foreign Policy," in *China Rising: Power and Motivation in Chinese Foreign Policy*, eds. Yong Deng and Fei-Ling Wang (Lanham, MD: Rowman & Littlefield Publishers, Inc., 2005), 73–102.

10 Camilla T.N. Sørensen, "Intensifying US-China Security Dilemma Dynamics Play Out in the Arctic: Implications for China's Arctic Strategy," in *Redefining Arctic Security: Arctic Yearbook 2019*, eds. Lassi Heininen, Heather Exner-Pirot and Justin Barnes (Akureyri, Iceland: Arctic Portal, 2019), 1–15, https://arcticyearbook.com/images/yearbook/2019/Scholarly-Papers/21_AY2019_Sorensen.pdf.

11 For a comprehensive discussion regarding the security dilemma concept see Shiping Tang, "The Security Dilemma: A Conceptual Analysis," *Security Studies* 18, no. 3 (2009): 587–623.

12 Amir Lupovici, "The Dual-Use Security Dilemma and the Social Construction of Insecurity," *Contemporary Security Policy* 42, no. 3 (2021): 261–262.

13 Ibid., 274.

14 Andro Mathewson, "Un-crewed Underwater Technologies in the Arctic," *Wavell Room*, July 28, 2021, https://wavellroom.com/2021/07/28/uncrewed-underwater-vehicles-artic/.

15 Margaret M. Pearson, Meg Rithmire and Kellee S. Tsai, "China's Party-State Capitalism and International Backlash: From Interdependence to Insecurity," *International Security* 47, no. 2 (2022): 135–176.

16 Ibid., 142–150.

17 Isaac B. Kardon and Wendy Leutert, "Pier Competitor: China's Power Position in Global Ports," *International Security* 46, no. 4 (2022): 43.

18 Adam Stepien et al., "China's Economic Presence in the Arctic: Realities, Expectations and Concerns," in *Chinese Policy and Presence in the Arctic*, eds. Timo Koivurova and Sanna Kopra (Leiden and Boston: Brill, 2020), 119.

19 Marcin Kaczmarski, "The Sino-Russian Relationship and the West," *Survival* 62, no. 2 (2020): 201.

20 Li Hui, "Sino-Russian Relations Embark on a New Journey" [中俄关系迈上新征程], *Interpret: China* (2023, original work published 2020), https://interpret.csis.org/translations/sino-russian-relations-embark-on-a-new-journey/.

21 Alexander Lukin, *China and Russia: The New Rapprochement* (Cambridge: Polity Press, 2018), xi.

22 Thomas des Garets Geddes, "The Art of Tasseography: China – Russia Relations as Viewed from China," *RUSI*, May 30, 2023, www.rusi.org/explore-our-research/publications/commentary/art-tasseography-china-russia-relations-viewed-china.

23 Alexander Gabuev (@AlexGabuev), "Fresh Chinese Customs Data for 5 Months of 2023 is Out Today. Sinification of Russia's Economy Is on Steroids . . . Again. CN-RU Trade in Jan-May +40.7% ($93.8b) RU exports to CN +20.4% ($50.85b) RU imports from CN +75.6% ($42.95) Source: http://customs.gov.cn/customs/302249/zfxxgk/2799825/302274/302275/5070327/index.html," *Twitter*, June 7, 2023, 8:02 a.m., https://twitter.com/AlexGabuev/status/1666326674974146561.

24 Atle Staalesen, "Arctic Shipping and Energy on Putin's Agenda with Xi Jinping," *The Barents Observer*, March 21, 2023, https://thebarentsobserver.com/en/industry-and-energy/2023/03/arctic-energy-agenda-xi-meets-russian-pm-mishustin.

25 Alexander Gabuev, "China's New Vassal: How the War in Ukraine Turned Moscow Into Beijing's Junior Partner," *Foreign Affairs*, August 9, 2022, www.foreignaffairs.com/china/chinas-new-vassal.

26 "Warning Against 'Great Fracture', Secretary-General Calls on General Assembly to Reconnect with Organization's Values, Uphold Human Rights, Restore Trust," *United Nations*, September 24, 2019, https://press.un.org/en/2019/sgsm19760.doc.htm.

27 Zhao Minghao, "Is the New Cold War Inevitable? Chinese Perspectives on US-China Strategic Competition," *The Chinese Journal of International Politics* (2019): 389. Not all agree with defining the US-China competition as a new Cold War. For example, Feng Yujun from Fudan University characterises this emerging international trend as "small divergence" defined as "an accelerated disruption and reorganization of the global supply and production chains brought about by the post-Cold War wave of globalization; continuous renewal and reworking of global trade and investment rules; and reform and reshaping of global and regional security architecture." Feng Yujun, "The Post-Russia-Ukraine Conflict World: Not a Splitting into Camps, But Rather a Small Divergence [俄乌冲突后的世界：不是"阵营化"，而是"小分流"]," *Interpret: China* (2023, original work published July 22, 2022), https://interpret.csis.org/translations/the-post-russia-ukraine-conflict-world-not-a-splitting-into-camps-but-rather-a-small-divergence/.

28 John J. Mearsheimer, "Bound to Fail: The Rise and Fall of the Liberal International Order," *International Security* 43, no. 4 (2019): 43–45.

29 Yan Xuetong, "Bipolar Rivalry in the Early Digital Age," *The Chinese Journal of International Politics* (2020): 334–340.

30 Rasmus Gjedssø Bertelsen, "Arctic Order(s) Under Sino-American Bipolarity," in *Global Arctic: An Introduction to the Multifaceted Dynamics of the Arctic*, eds. Matthias Finger and Gunnar Rekvig (Cham, Switzerland: Springer, 2022), 463–481.

31 Torbjørn Pedersen, "The Politics of Research Presence in Svalbard," *The Polar Journal* 11, no. 2 (2021): 418.

32 Ibid., 419.

33 Astri Edvardsen, "Russia Wants to Cooperate with BRICS Countries on Research on Svalbard," *High North News*, April 14, 2023, www.highnorthnews.com/en/russia-wants-cooperate-brics-countries-research-svalbard.

34 Atle Staalesen, "Lavrov: Fate of Arctic Council Is at Stake," *The Barents Observer*, May 12, 2023, https://thebarentsobserver.com/en/arctic/2023/05/lavrov-fate-arctic-council-stake.

35 Melody Schreiber, "China Will Not Recognize an Arctic Council Without Russia, Envoy Says," *Arctic Today*, October 17, 2022, www.arctictoday.com/china-will-not-recognize-an-arctic-council-without-russia-envoy-says/?wallit_nosession=1.

36 Du Debin and Ma Yahua, "一带一路：中华民族复兴的地缘大战略" [One Belt and One Road: The Grand Geo-Strategy of China's Rise], 地理研究 [*Geographical Research*] 34, no. 6 (2015): 1005–1014.

37 Alexander Lukin and Dmitry Novikov, "Sino-Russian Rapprochement and Greater Eurasia: From Geopolitical Pole to International Society," *Journal of Eurasian Studies* 12, no. 1 (2021): 28–45.

38 Barry R. Posen, "Command of the Commons: The Military Foundation of US Hegemony," *International Security* 28, no. 1 (2003): 5–46.

Index

For Product Safety Concerns and Information please contact our EU
representative GPSR@taylorandfrancis.com
Taylor & Francis Verlag GmbH, Kaufingerstraße 24, 80331 München, Germany

www.ingramcontent.com/pod-product-compliance
Lightning Source LLC
Chambersburg PA
CBHW060310220326
41598CB00027B/4292